# Sailing the South Seas
## A 15-Year Adventure in Shaula

Alice Dews

Sailing the South Seas:

A Fifteen-Year Adventure in *Shaula*

Copyright 2016 by Alice Dews

First printing May 2016

All rights reserved

No part of this book may be used or reprinted in any matter whatsoever without written permission

Cover photo [*Shaula* leaving Bora Bora] by Debbie Morley

All other photos are from the author's collection

Maps by Dan Dews

Printed in the United States of America

ISBN-10:1533195560

ISBN-13:978-1533195562

CONTENTS

| | | |
|---|---|---|
| Chapter 1 | Sailing in Calm Waters | 1 |
| Chapter 2 | Offshore Cruising | 13 |
| Chapter 3 | California and Mexico | 31 |
| Chapter 4 | French Polynesia | 51 |
| Chapter 5 | Hawaii | 77 |
| Chapter 6 | Sea of Cortez | 109 |
| Chapter 7 | Mexico to New Zealand | 123 |
| Chapter 8 | New Zealand to Australia | 191 |
| Chapter 9 | Australia | 207 |
| Chapter 10 | Vanuatu and New Caledonia | 269 |
| Chapter 11 | Exploring Australia | 315 |
| Chapter 12 | Returning Home | 337 |

# PROLOGUE

The first night after leaving Neah Bay for Hawaii was the worst sailing I had ever experienced. It was pitch dark, rain was pelting down, and wind was whistling through the rigging. The weather report for ocean waters was 20-to-25-knot winds and five-to-nine-foot seas. Due to the wild gyrations of the boat as it barreled through the steep waves we were constantly thrown around. The motion was horrible. I held on with a tight grip to the grab rail on the cabin top to keep from sliding off the cockpit seat.

Dressed in heavy slacks, a long sleeve shirt, a sweater and a winter jacket, with full length foul weather pants and jacket on top, I was cold, wet, thirsty, hungry, but most of all terrified. It was my watch, which meant I had to climb up the stairs to the cockpit every 10 minutes to scan the horizon 360 degrees to look for ships. The boat was rocking and pitching wildly as large waves crashed into the hull. Looking forward all I could see was a black ocean. Dan was asleep in his bunk for his three-hour off-watch. I was grimly standing my watch, worried about striking unseen objects in our path, such as debris, logs, or whales.

What I thought would be a pleasant sailing adventure, spending the summer sailing to Hawaii and back, had turned into a nightmare. I wanted to yell "Stop the boat, I want to get off!" Unfortunately, there was no turning back. Once we were out on the ocean with the strong winds behind us, we couldn't turn around. Sailing into the wind would be far worse than sailing downwind. We were committed to a 3,000-mile voyage. I wondered how sailing, my favorite activity for the last 25 years, had turned into this. Why had I agreed to sail to Hawaii?

# 1 SAILING IN CALM WATERS

## Learning to Sail

My love affair with sailing began when I was in my mid-20s. I was living in Berkeley, California, and working as a chemist at the U.S. Department of Agriculture. My roommate and I decided to take a sailing course at the small lagoon known as Albany Pond. It was a small class of eight students taught by John C. Beery. We were given classroom instruction on the nautical terms for a sailboat and its rigging and the principles of sailing. Each student was then given the use of an eight-foot El Toro sailboat to sail around the pond. It was a delightful Saturday afternoon activity practicing tacking, jibing, and going into irons in the gentle sea breezes. In the last part of each class we raced against each other in order to hone our skills. I seldom won a race, but I loved the sensation of sailing the boat across the water using the wind for propulsion.

After the course was over I joined the University of California Sailing Club. The sailing club had a fleet of fiberglass sailboats known as Lido 14s that members could sail around Berkeley Harbor. It was easy to sail the larger Lido 14s after learning the basic principles of sailing on the smaller El Toros. In the mild weather of the Bay Area it was possible to sail year round. For the next four years I spent most weekends sailing around Berkeley Harbor and occasionally took all-day trips to Angel Island with groups of six or eight sailors on the 25-foot sailboats belonging to the club. When sailing was offered as a physical education course for U.C. students I volunteered as an instructor. As instructors we took two or three students at a time out in the harbor for one-hour classes of hands-on sailing. It was great sharing my enthusiasm for sailing with students who were just learning to sail.

Four years after taking up sailing I moved to Eugene, Oregon to attend graduate school at University of Oregon. It was difficult to leave the Bay area where I had easy access to sailing any time at Berkeley Harbor, only a short distance from home. In addition, I found the cold, rainy fall and winter in Eugene depressing after living for seven years in Berkeley. My first winter in Eugene I met Dan, a Biology graduate student at the university. Dan wasn't a sailor, but when I told him how much I missed sailing, he agreed to go with me to Portland in the spring to look for a used sailboat. I found what I was looking for right away—a 10-foot fiberglass sloop, large enough for two people to sail, but small enough that it would fit on top of the car.

That spring and summer Dan and I spent many weekends driving out to Fern Ridge Reservoir, transporting the sailboat and mast on a cartop carrier. We lifted the boat off the car at the boat-launching ramp, rigged it with the mast, boom, and sails, and sailed for hours around the lake. It took only a few lessons to teach Dan how to sail; he was a quick learner. Soon he became as avid a sailor as I was. Sailing became our favorite pastime. After our sails we often drove to one of the U-pick farms that surrounded the reservoir to pick fruits or vegetables in season. We returned to Eugene with large bags of blueberries, gooseberries, or walnuts at 10 or 12 cents a pound. I could endure the miserable fall and winter weather knowing that spring would arrive soon and we would spend warm, sunny afternoons sailing.

We saw a few other boats at Fern Ridge Reservoir, but one in particular caught our attention. It was a wooden strip-planked sailboat, 12 feet long, with foredeck made of alternating mahogany and Douglas fir strips, covered with clear fiberglass to protect the wood. We asked the owner where he bought it, and he told us about

Van Doorn, a Dutch cabinetmaker who built wooden boats in a shed at his house. We drove out to Van Doorn's home on the outskirts of Eugene, and saw the identical boat that he had just completed for sale for $1,200. When we got married at the end of May we decided to buy this beautiful wooden sloop as a wedding present for ourselves. We pooled our savings of $600 each and drove out to make our purchase. The boat came with a trailer we pulled behind our Volkswagen Beetle, and it just fit in the garage at the house we were renting. We loved our new boat, and spent as much time as we could sailing on the reservoir. The boat was large enough our friends could come along. After several summers we joined Bob and Linda and their young son Ben for an adventurous one-week cruise in Washington's San Juan Islands in their 25-foot sailboat. The next summer, after they sold their boat, we joined them for a one-week charter cruise from Seattle to the San Juan Islands on a wonderful black-hulled wooden boat called *Midnight Sun*.

During our years of attending graduate school in Eugene, Dan and I read many books about sailing adventures across the Pacific and around the world. Dan was intrigued with the concept of being able to sail wherever you wanted and being entirely self-reliant. I liked the idea of travel to distant countries, but the stories of sailboats being boarded by pirates, battered by storms, sunk by whales, and pitchpoled while rounding Cape Horn were much too frightening for me. Ocean sailing seemed too dangerous. Local sailing on small lakes and on Puget Sound was what I enjoyed, but Dan was not satisfied with local sailing. Many adventures lay ahead.

## How I Became a Cruising Sailor

When Dan and I were married in May 1968, we took the money we had received for wedding presents to REI in Seattle to buy sleeping bags, foam mattresses, and Kelty backpacks. Our plan was to take a 10-day backpacking trip on a section of the John Muir trail in Sequoia

National Park. We chose to go in early September because the weather was cooler than in mid-summer and the trails were less crowded after the start of school. This would be a new experience for both of us. Dan had taken a backpacking trip with a friend when he was in high school. They hiked for a week in the San Gabriel Mountains near his home in Southern California. My experience was mostly day hiking, which I had done since I was a child, first in the Smoky Mountains near Knoxville, Tennessee, and later in the hills near Berkeley. My only experience with overnight camping was a one-week burro trip in the Sierra Nevada mountain range about five years earlier with a group of hikers in the Sierra Club. We hiked along the trail all day, carrying only a light knapsack for water and snacks, while burros carried our bedding and clothing and all the food and cooking utensils. Dan and I both loved hiking in the mountains and I was game to try backpacking.

It took weeks of planning to pick out a section of the John Muir trail that we could hike in 10 days. We made long lists of equipment and supplies to take in our backpacks. After Labor Day we packed all of our food and equipment into backpacks, and stuffed them into our Volkswagen Beetle. It was a two-day drive down the freeway from Eugene to Sequoia National Park.

Even with careful planning, whittling down our supplies to only the essentials, we still had heavy packs—Dan carried 60 pounds and I carried 40 pounds. It was a struggle as we headed up the mountain, starting at 9,000 feet and climbing up to the pass at 11,000 feet. The first few days were exhausting. Our packs weighed heavily on our shoulders and our new boots rubbed raw spots on our feet. When we reached our campsite each evening we had to gather kindling and start a campfire to cook our mostly freeze-dried dinners. We ate the meager fare with our only utensils, a Sierra Club cup and a spoon-fork combination for each of us. Nevertheless, we loved the fresh, clear air in the mountains, the fantastic scenery, and the exhilaration of reaching the top of the passes and realizing we had accomplished this all under our own power. About half way through the trip the hiking became much easier, and by the end we felt we could go on forever if only we had more food.

In the two following years we took longer backpacking trips of two weeks, then three weeks, each time hiking different sections of the John Muir trail. Three weeks was our limit, because we couldn't carry any more food and still keep our packs at a manageable weight. On our three-week trip, we felt marvelously fit towards the end, but we were literally starving. We were expending more calories than we were taking in, and both lost five-to-10 pounds. For the last few days we had vivid dreams of indulging in wonderful hot fudge sundaes, while in reality all we had to eat were snacks of dried fruit and nuts, crackers, and freeze-dried dinners such as pork chops that tasted like cardboard.

Backpacking was the perfect preparation for cruising because it made us appreciate all the luxuries of a small boat. Friends we met, who had sold their homes and moved onto a cruising boat, often complained about the discomforts of a small boat. They missed the convenience of refrigerators, freezers, microwaves, and washing machines. We found cruising has many of the advantages of backpacking without the discomforts. No longer did we have aching backs and sore feet from hiking. No longer did we have to collect water from streams, light a campfire to cook our dinner, and subsist on the meager fare that we could carry on our backs. No longer did we have to crawl in our sleeping bags after dinner for warmth. While cruising we were carried effortlessly over the water by the wind, at a speed greater than we could hike, and we saw miles and miles of beautiful scenery. At the end of the day we could anchor in a quiet bay, have a leisurely dinner of fresh food, and listen to the radio or read all evening without ever feeling exhausted. We became confirmed cruisers and never had the desire to go backpacking again.

## Coastal Cruising in Australia

In 1971 our backpacking ended abruptly when Dan accepted a teaching job in New South Wales, Australia. Dan had received his teaching certificate and a master's degree in Biology. We both became PhD candidates in Biology, taking courses October through June, and working on research projects year round. When we were both within about a year of finishing PhDs in biology, we became disillusioned with research, and realized that jobs for PhDs were scarce. Neither of us wanted to become a college professor. Dan wanted to teach at the high school level, and I wanted to work in a lab. Dan began interviewing for jobs in the Pacific Northwest. Unfortunately, Boeing had just lost the Super Sonic Transport contract. It laid off massive numbers of engineers, resulting in a decline in school enrollment in the Seattle area. Teaching jobs were scarce, especially for beginning teachers. After travelling for interviews as far south as Brookings, Oregon, east to Walla Walla, Washington and north to Vancouver, British Columbia, with no job offers, Dan signed up for an interview with a recruiter from the Department of Education in New South Wales. He was hired immediately on a two-year contract.

We sold our sailboat, furniture and other items, packed the rest of our belongings in our Volkswagen Beetle, and drove to Glendale, California, where Dan's parents lived. We put our belongings in storage, left the Volkswagen for Dan's father to sell, and flew to Sydney in September. Dan was assigned to teach science at a high school in the suburb of Kirrawee. I found a job taking telephone ads at the Sydney Morning Herald, as there were no jobs for women in my field.

Australia has no high mountains, nor an equivalent of the John Muir trail, but it is surrounded by ocean, and the coast is dotted with harbors. We rented a small apartment in Cronulla, south of Sydney, near Gunnamatta Bay where we could moor a boat. From our apartment we had a peek-a-boo view of the Pacific Ocean.

Living in Australia seemed the perfect opportunity to practice sailing offshore along the East Coast of Australia. All we needed was a suitable, sturdy boat for ocean sailing. It took us only a few months to buy a used fiberglass sailboat. It was 23 feet long, designed for racing rather than cruising, but it had two comfortable berths, an alcohol cooking stove, and toilet. The only navigational equipment it had was a knot meter and a compass. It had an outboard motor that we mounted in the cockpit to use when the wind died. There was no standing headroom, so we had to crouch when moving around inside the boat, and to prepare and cook meals while sitting down. Nevertheless it was excellent at sailing, strong and seaworthy. We moored the boat in the small harbor in Gunnamatta Bay, just a 10-minute walk from our apartment. From there we sailed out to the coast to the many harbors and anchorages around the Sydney area. On the long school vacations we sailed about 100 miles north to Port Stephens and about 100 miles south to Jervis Bay. We stopped at all the small harbors along the way, sometimes being the only sailboat among a fleet of fishing boats. Sailing in Australia was an invaluable experience in learning how to handle a sailboat in the ocean, navigate along the coast and anchor in bays, harbors, and inlets.

## Cruising to Desolation Sound

At the end of Dan's two-year teaching contract in New South Wales we sold our small sailboat and returned to the States. We searched for a place to live where we could find jobs and sail, and settled on Seattle because it is surrounded by saltwater and has easy access to the San Juan Islands and the waters of British Columbia. Dan got a job as a teacher in Seattle Public schools, and I was hired as a Research Technician at University of Washington. Our life in Seattle was similar to our graduate student life. To save money we rented a basement apartment, and we bought our furniture at garage sales or

Goodwill. We shared one car. Within a year of settling down we found a used sailboat for about $20,000. With our frugal lifestyle and both of us working we had saved enough money to buy it with cash. It was a 30-foot fiberglass Rawson with a diesel inboard engine and standing headroom. Dan built a mahogany drop-leaf table for the main cabin and took out the poorly built cabinets in the forepeak. Now we had all the comforts of home—two comfortable berths with thick mattresses, kerosene cooking stove, a wood-burning heating stove, a toilet, and room to store all the clothing, books and food we needed for a summer of cruising. Dan had summers off from his teaching job and I took leave without pay from my job so we could spend two to three months each summer on our boat. Our favorite cruising area was Desolation Sound, in British Columbia, about 240 miles north of Seattle. We learned to jig for rockfish, dig clams and gather oysters to supplement our canned and dry food. In Desolation Sound there was a small grocery store at Squirrel Cove on Cortes Island, and another one at Heriot Bay on Quadra Island where we re-supplied with fresh fruits and vegetables throughout the summer. We discovered many trails on the islands for hiking and lakes for swimming. It was a healthy existence. We enjoyed the quiet of sailing in the flat waters of Puget Sound and north away from traffic noise, the only sound while sailing being the slap of waves and the swoosh of water against the hull. As in backpacking, we delighted in the clear skies and the view of the moon and stars.

## Buying an Ocean Cruising Sailboat

Shortly after moving to Seattle we joined the Puget Sound Cruising Club. This was an informal club of people who liked to sail, many of whom had desires to sail across the ocean. Meetings were held once a month at which members gave talks and slide shows of cruises they had taken to exotic places such as Hawaii, South America, French Polynesia, Europe, Australia, or New Zealand. The talks were inspiring, stimulating our desire to do some long-distance cruising.

The Rawson was roomy and well suited for sailing in protected waters of Puget Sound, and north to Desolation Sound. However, for ocean sailing in strong winds and big seas it had problems. It had a tendency to hobbyhorse in any sort of chop, and was difficult to steer in strong winds because of its weather helm. When sailing downwind it took a lot of force on the tiller to keep the boat from rounding up into the wind.

During the five years that we owned the Rawson we were constantly looking for a suitable offshore boat. After many long searches, attending boat shows, and reading advertisements in sailing magazines, we found what we thought was the perfect boat—a 28-foot Bristol Channel Cutter (BCC). After taking a test sail on Puget Sound with a BCC owner we were convinced this was the boat for us. It had everything that the Rawson lacked. Used ones were scarce. We went as far as Port Hardy in British Columbia to look at one, but being an early version it had some design flaws that would be costly to correct. However, we were not deterred—we decided to buy a new BCC.

The Bristol Channel Cutter was made in Costa Mesa, California, by the Sam Morse Company. Sam Morse was unique among boat builders. He allowed the customer to purchase a boat at any stage, from just a hull and deck to a fully finished boat. Because we were in our late 40s, and had no boat-building skills, we didn't want to spend 10 years finishing a boat in our spare time. We ordered a boat with the interior furniture completed, but nothing painted or varnished. The wiring, water tanks, and fuel tanks were installed, but Dan would install the lamps, the sink, the heating stove, and many other jobs he could do easily. By electing to do some of the work ourselves we got the price down to $60,000. It was wonderful to be able to buy a new custom-made boat. I could tell from the stars in Dan's eyes that we would be sailing this boat across the Pacific.

## Arrival of our Ocean Cruising Boat

Our new sailboat arrived by truck from Costa Mesa, California, in October 1981, eight months after we ordered it. We were thrilled to see it arrive at Shilshole Bay Marina. It had a shiny ivory hull with dark green cove stripes, mahogany bulwarks and hatches, a wooden bowsprit and boomkin, and a bright yellow mast and boom. It was a semi-custom boat in that we had asked for features that were not standard. We had ordered an arrangement of berths in the cabin so we could install a heating stove, boom gallows (a high horizontal beam in the cockpit area which supports the boom when not in use), and taller stanchions (stainless steel supports at the edge of the deck through which the lifelines run. The lifelines keep sailors from falling overboard). We had a quarter berth for sleeping a third person and a cooking stove with an oven.

The day it arrived we arranged to have the boat lifted from the truck by travel lift and placed on stands in the yard at Shilshole where we could work on it. We varnished all the exterior wood, installed the bowsprit and lifelines, cleaned the wax off the hull then applied two coats of anti-fouling paint. After a month of work we ordered two cases of champagne and invited our friends to a launching party at the marina. Our niece Ann broke a bottle of champagne against the bow and we christened the boat *Shaula*. (Shaula is a navigation star in the constellation Scorpio; we chose it because we liked the sound of the name.)

Once our boat was in its slip at Shilshole there was enough work to do to keep us busy all winter and spring. Dan installed lights, kerosene stove, and plumbing for the water tanks. I spent evenings painting and varnishing the interior. We had paid a third down when we ordered it, a third during the building of the boat, and the last third at delivery. By this time we had sold our Rawson for $25,000, enough for the one of the payments. The rest came out of our savings. By June we were ready to sail north for the summer.

Our new boat was everything we could want. It was a cutter, with a main, jib and staysail, making it more flexible for sailing in any kind of wind. In light winds we sailed with all three sails, while in strong winds we could sail with only a reefed mainsail and staysail. It was both seaworthy and sea kindly. We spent delightful summers sailing to Desolation Sound and Princess Louisa Inlet. For part of each summer we took along our niece Ann, or Zach, the son of friends from Eugene. Having enthusiastic young people aboard to share in sailing, hiking, swimming and fishing was delightful, and added greatly to the enjoyment of our summer cruising.

# 2 OFFSHORE CRUISING

## Safety at Sea

When we decided to sail offshore, one of our primary concerns was safety. Some of the safety gear recommended for ocean crossing is expensive, but when your life is involved it seems foolish to skimp on the equipment. Our biggest expenditure was for a life raft. We bought the valise-style, which we keep in the lazarette. If our boat were taking on water and in danger of sinking, we could lift the life raft out of the lazarette, tie the towrope to a cleat on the stern of the boat, and throw it overboard. The life raft automatically inflates when it hits the water, and we could climb into it, and cut the line as our boat was sinking. This scenario seemed pretty scary, but of course we hoped it would never happen.

Another expensive piece of equipment we installed was a self-steering vane, which steers the boat in relation to the wind. Hand steering for hours on end is fatiguing, so our self-steering vane is an important safety device. It's better than having a third crew member because it can keep a course in heavy weather without complaining and uses no power. All we have to do while on watch is check the compass to see if the boat is on course, and bring it back on course by pulling a string attached to the gear on the self-steering vane. Each pull makes one click of the gear and changes the course by six degrees. Dan also installed an electric autopilot connected to the tiller for when the winds are too light for the self-steering vane to work. With the autopilot we can change course either one degree or 10 degrees by pushing buttons on the control panel. The other important safety equipment we installed when we first got the boat is our ham radio. It can transmit and receive over thousands of miles. We use it to communicate with people on other boats and on land, and to receive weather forecasts. It can also be used in an emergency in case the boat is damaged or either of us injured. Like everyone we have a VHF radio, used to communicate with nearby boats or ships, or to call the Coast Guard in an emergency. However it's relatively short range compared to the ham radio. Other safety equipment we bought was an EPIRB—an emergency position indicating radio beacon. In case of an accident on board we could turn on the EPIRB which sends a signal to orbiting satellites. Land stations use the signal to identify and locate us. The Coast Guard would then alert any ship in the vicinity to come to our aid.

Out on the ocean we wear safety harnesses whenever we leave the cabin, clipping our tethers to a sturdy metal strap bolted at the bridge deck in the cockpit, or onto a jack line, which runs the length of the deck on either side of the boat. Hopefully this prevents one of the biggest dangers of sailing offshore: falling overboard. We also installed a Lifesling, a horseshoe-shaped flotation device designed to rescue a person who falls overboard.

Before we left on our first passage we took several courses on boating safety. We took advantage of the power squadron course on basic seamanship to learn the rules of the road, docking, anchoring, and recognizing the different types of ships. Through our jobs we both took first aid classes and CPR training. We took an advanced first aid class intended especially for fishermen. They taught us how to splint broken bones, bandage wounds, and how to evaluate symptoms and treat serious illnesses. Most memorable was the instruction for giving CPR on board. When one of the classmates asked how long to continue CPR on board, we were told as long as you can. This was discouraging—with only two people on board, the person performing CPR could become exhausted. If the CPR didn't succeed there was little hope of a good outcome.

The most extensive course we took was an all-day Safety at Sea workshop at Magnuson Park in Seattle. We had morning and afternoon sessions on all aspects of safety. The highlight of the afternoon was a demonstration at the swimming pool of how to launch and climb into a life raft. After the life raft was launched participants of the workshop could jump into the water, swim to the life raft and board it. The instructors used a nearly new life raft, but one that had been recently re-packed. After tying the line to a rail at the end of the pool, one instructor threw the life raft into the water, but nothing happened. With much effort he pulled repeatedly on the line, but the life raft refused to open. The instructor concluded that something went wrong in the re-packing, and the demonstration had to be cancelled. We were shocked this ultimate life-saving equipment could fail.

The Lifesling was developed by a group of Seattle sailors as a means of retrieving a person if they fell overboard. A sailing couple at Shilshole Bay Marina offered to take us and another couple and their teenage son out in their boat along with an instructor for a practice session on using the Lifesling. The instructor explained how to toss the Lifesling overboard on a 50-foot line, circle the boat around and into the wind so the Lifesling was close to the victim. The victim could grab it and put it around his chest. Then the sailor would pull him up to the boat and hoist him aboard with a pulley attached to the boom. To practice all we needed was a person to jump overboard, and the six of us would take turns retrieving the victim. The teenage son volunteered, after his mother agreed to buy him a CD of his favorite rock band. He donned his mother's wet suit and bravely jumped in, initially sinking so his head was underwater. The water was cold, and having a wet head added to his discomfort. Unfortunately, the first person to try to rescue the victim misjudged and the Lifesling ended up too far away, so he had to repeat the procedure. However by this time the victim's long immersion in the cold water caused his lips to turn blue and his teeth to chatter. The instructor was anxious to have all six of us try the procedure. But when we saw that even after drying off and being wrapped in a blanket the poor teenager was still shivering, none of us wanted him to jump in again. We had learned that a Lifesling could save lives, but we weren't willing to have the victim suffer from hypothermia to prove it, so we called it a day and returned to the marina. It was obvious that a Lifesling wasn't easy to use even in the calm waters of Puget Sound, and would be extremely difficult out on the ocean in strong winds and big seas.

From all of our safety courses we learned the best way to reduce the risks of offshore sailing is to follow safe practices and not rely on safety equipment that can fail. This meant we would never leave the cabin without using our safety harnesses, we would never enter a harbor at night, we would keep a careful lookout for ships to avoid a collision, we would change course to avoid collision with a ship even though a sailboat has the right of way, and we would try to get enough rest or sleep during a passage that we were not fatigued. We knew that making an ocean passage could never be completely safe, but we could reduce the risks to a reasonable level. Dan always claimed it is safer crossing the ocean in our small boat than driving on the freeway, and I believe this to be true. Everything in life is risky, but we were willing to take the small risks involved in crossing the ocean in our small sailboat.

## Shakedown Cruise to Hawaii

After we sailed for three summers in local waters in our new boat, Dan convinced me to sail *Shaula* to Hawaii and back during the summer of 1985. It would be a shakedown cruise to test the boat and ourselves. I was apprehensive about the trip because I suffered from seasickness in ocean swell and had a fear of whales. Nonetheless, knowing that Dan had had the dream of sailing offshore for many years, I agreed to go. We had learned Morse code to get our amateur radio licenses four years before and had taught ourselves to navigate by sextant, although our positions were sometimes off as much as 10 miles. Dan installed a self-steering vane so we wouldn't have to hand-steer all the time. I made canvas lee cloths to keep us from falling out of our berths, and a cockpit awning to protect us from the tropical sun.

The four-day trip in calm water out to Neah Bay, Washington, was uneventful. This was the jumping off place for Hawaii. Once we left shore we would be out of sight of land until we reached Hawaii. It blew a gale shortly after we departed Neah Bay, and the motion was horrible. The boat lurched and plunged in the large ocean waves. We were cold, wet, and miserable as we sailed southwest along the Washington and Oregon coasts. I was terrified, particularly after the sun set when we were sailing pell-mell into pitch darkness. I could imagine all kinds of floating objects we might hit, such as logs, debris, containers from ships, or whales. We stood three-hour watches, which meant struggling into foul weather overalls and jacket over several layers of clothing at the beginning of each watch. During the watch we climbed into the cockpit every 10 minutes to look around 360 degrees for ships, checked our course, and adjusted the self-steering vane if the wind had shifted. We recorded the speed and course in the ship's log every 30 minutes. This routine was relieved only by sitting on the cabin floor next to the ladder reading a book by flashlight, waiting for the 10-minute timer to go off for the next climb into the cockpit. The three hours seemed interminable. At the end of my watch I would wake Dan and struggle out of my wet foul weather gear. I'd collapse on my berth, attach the lee cloth, and try to sleep, listening to the rush of water against the hull right next to my ears as the boat crashed through the waves. A few days after we departed I came down with a cold, which developed into a hacking cough, making it difficult to sleep on my three-hour off-watches.

Dan and his father had scheduled daily contacts on ham radio based on the times reception would be good. They used Morse code for their conversations. The dots and dashes of Morse code could be heard above the noise, while their efforts to use voice communication usually failed with the inadequate antennas they were using. Unfortunately, communication ended on the ninth day out when our ham radio suddenly quit transmitting in the middle of a conversation. Although ham radios often go out due to saltwater corrosion on a boat, we were afraid Dan's father would imagine the worst—that our boat had sunk in the middle of the ocean. By this time we were both exhausted, and standing watches had become a real chore. On the evening of the 10th day out Dan said, "This trip has cured me of the desire to cruise." I agreed wholeheartedly.

As we got further south the wind and seas decreased and the motion became much more tolerable. I could cook simple meals, although I frequently cursed at the sea gods when the seas seemed to increase dramatically around five in the afternoon, just when I was trying to get cans out from under my berth to heat up for dinner.

We were taking sun sights at least twice a day to determine our position. This involved 30 minutes or more of calculation for each sight, which we would both work until our calculations agreed. Although this took time away from our reading, it gave us a sense of accomplishment that we could determine where we were with only a sextant and a set of calculations. It was immensely satisfying to plot our noon positions on the chart every day, and join the dots to show our path across the ocean.

After two weeks at sea we finally started getting northeast trade winds. With the wind behinds us the motion was more comfortable, and we could read books more easily, and listen to Prairie Home Companion tapes. With the improvement in conditions our moods also improved. The weather was getting warmer, and we began to see flying fish soaring over the waves. Now we were really beginning to enjoy the cruise. One afternoon we were excited to see on the horizon a large four-masted sailing vessel. We called them on VHF radio, and learned that it was a Japanese training ship, the *Nippon Maru*, with 250 cadets on board. They were sailing from Hilo, Hawaii to Long Beach, California. It was dark by the time we were close, but we could see their red and green bow lights as we crossed paths

On the 21st day Dan trailed out our long fishing line with a large lure. We began seeing white birds with a long yellow tail feather soaring high above our boat. These were birds we had never seen before—tropicbirds. One began hovering above our boat eyeing our lure. It occasionally dove straight down towards the lure, but flew off when it got close and realized it was not edible. A couple of days after putting out our lure Dan landed the largest fish we've ever caught—an 18-pound, 31-inch tuna. We had a couple of delicious fish dinners.

On the 23rd day, with only 180 miles to go, we were getting excited at making our first landfall. The next day we calculated our position to be about 60 miles north of Hilo. Because of the height of Mauna Kea, 13,000 feet, the island should have been easily visible from this distance, but all we could see was a huge towering cloud ahead of us. By this time we were motoring because the winds had died. Dan mentioned casually that our calculated position could be wrong, and we might have sailed past the Hawaiian Islands. This seemed plausible to me, given our inexperience at navigating by sextant. I panicked, and insisted Dan call the Coast Guard in Hilo on VHF radio to find out if they could determine our position. The man at the Coast Guard who answered Dan's call said they had no way of doing that, but we could tell by the loudness of his voice that we must be fairly close. That was very reassuring.

Then suddenly in mid-afternoon the giant cloud parted and we could see the top of Mauna Kea and beaches and pineapple fields at the shoreline. What a beautiful sight. Dan shouted "Land ho!" We had made our first landfall. We were exuberant. I had never been so glad to see land. However, my joy was dampened by knowing we would have to leave in a few weeks to sail back to Seattle. I was already dreading the trip back.

## Entering Hilo Harbor

After making landfall we still had to motor for a few hours to enter Hilo Harbor. Dan wanted to anchor in Reed's Bay, a small bay outside the entrance to Radio Bay, while I wanted to continue on into the wharf at Radio Bay so I could set foot on land. After a short argument, the first disagreement we had had during the whole passage, I agreed to anchor overnight. However, when we dropped the anchor in Reed's Bay and backed up to set it, we found the anchor dragged across the bottom, which appeared to be smooth lava covered by a thin layer of sand. Thus I won the argument, and we proceeded to Radio Bay. By the time we entered the bay it was 7 pm, and the sun was about to set. A dozen or more sailboats were tied to the long wharf with Med ties. (Boats were anchored well out with their bow or stern tied to the wharf so they were lined up perpendicular to the wharf rather than parallel. This is a common method of mooring in the Mediterranean, where wharf space is limited, hence the name Med tie). The harbormaster motioned with his flashlight for us to back into the first empty slot. Having never used a Med tie before we dropped the anchor and tried to back into the small slot between two boats. This is tricky because sailboats do not back straight, but tend to drift either to the right of left, depending on the pitch of the propeller. We couldn't back straight enough to get into the first or second spaces the harbormaster directed us to. He finally indicated an extra wide space further along the wharf and we were able to back in and toss our stern lines to a waiting sailor from another boat. How embarrassing to be so incompetent at mooring in front of all these sailors! However, we were rewarded when another sailor handed us two cans of cold Hawaiian fruit juice—a very welcome gift, as we hadn't had anything fresh for a couple of weeks.

We went to bed around 10 pm hoping to make up for lost sleep during the passage, only to be rudely awakened by the loud blast of the horn from a huge cruise ship tied up at a nearby wharf. We drifted back asleep around 11 pm, but at 2 am I woke up, climbed into the cockpit and looked around 360 degrees. My body was adjusted to the three hours on/three hours off watch schedule and I was incapable of sleeping through the night. It was so calm and beautiful to see the lights on shore and the boats tied up to the wharf and I was overjoyed not to be rocking around wildly in the big seas.

The next morning Dan called his parents in Glendale, California, to tell them we had arrived safely. It had been two weeks since Dan's last ham radio conversation with his father, and his parents were naturally concerned whether we were all right during that time. Dan's parents and brother planned to meet us on the Big Island of Hawaii. They were leaving in a couple of days for a one-week stay in Oahu and a second week on the Big Island to celebrate their 50$^{th}$ wedding anniversary. Dan asked if his father could purchase a new ham radio in California and bring it with them. We would reimburse them for it. It would be impossible to have our ham radio repaired during our short stay in Hawaii. With the new one installed we could communicate on the way back to Seattle. Once we got back to Seattle we could have the old one repaired and use it as a spare.

The third day after arriving I still had a slight cough and a sore throat, and was having trouble sleeping more than three hours at a time, so we travelled by bus to the Hilo Hospital emergency room for help. I wanted to be well for the passage back to Seattle. Imagine my surprise when the emergency room doctor told me he had sailed to Hawaii from San Francisco several years before in a Westsail 32. He understood the problems of passage making in a small boat, and was very sympathetic. He took a throat culture to rule out Strep throat, prescribed a mild tranquilizer to help me sleep while in harbor, and recommended I take a shot of rum at the end of my three-hour watches during the passage back to Seattle. The rum would make me sleepy, but would wear off in time for my next watch. Although we never drank alcohol while underway, I thought this was splendid advice, and would make our trip back much more pleasant.

I loved the fact that we were immediately accepted into the cruising community by virtue of the fact we had made an ocean passage. All the cruisers were extremely friendly, and enjoyed hearing about our passage, as we enjoyed hearing about theirs. Often we were invited aboard other boats for coffee and a visit. The camaraderie came about not just because we had all sailed across the ocean to get here, but because we shared a background of years of finding a seaworthy boat, preparing the boat for ocean cruising, and learning what we needed to know about navigation, weather, provisioning, and sailing in ocean waters in order to have a successful voyage. Having spent weeks crossing the ocean with no one to talk to during our watches, everyone seemed eager to socialize. We met Americans, Canadians, Italians, French, Swiss, Australians and New Zealanders. They had made passages from the Marquesas, Tahiti, the U.S. West Coast, or Mexico, and had fascinating stories to tell. During our stay in Hilo 10 more boats arrived and four departed. It was a true crossroads.

Some cruisers we met had made many passages, over a period of months or years. Others, like us, had just made their first passage. We felt sorry for a young couple who had sailed from Southern California with their seven-year-old daughter. All three were seasick the whole trip. They declared it was their first and last passage. We were impressed with a couple who had sailed from Australia, stopping at many islands along the way, and a 63-year-old sailor from Vashon Island who had built his own boat and had been cruising for 10 years. He was headed for American Samoa. Compared to most of the sailors we met we felt like complete novices. We still had much to learn.

## The Approach of Hurricane Ignacio

A week after our arrival in Hilo we learned that hurricane Ignacio had formed off the coast of Mexico and was on track to hit the Big Island in a few days. When Dan called his family at their hotel on Oahu, they were worried about the hurricane. They had been watching the weather reports on television. This is a cruising sailor's nightmare—having the boat damaged by a hurricane after reaching a safe harbor. All the boats stayed in the harbor waiting for the hurricane to arrive or pass by. The harbormaster started posting warnings outside his office showing the track of the hurricane. He came down to the wharf to tell the cruisers it would be wise to put a second anchor out further from the wharf and pull their boats at least 10 feet from the concrete wharf. If the hurricane hit Hilo, waves would come over the breakwater and the resulting surge in the harbor could smash boats into the wharf. Although our boat was the smallest in the harbor we had a sturdy fiberglass dinghy that is more suitable for carrying a heavy anchor than inflatable dinghies that most boats had. Dan and another cruiser volunteered to take people's second anchors in our dinghy and row them out towards the middle of the harbor to drop them so the boat owners could pull their boats well away from the wharf.

It was a frightening time. The approaching hurricane was on our minds constantly. Everyone listened to weather updates on VHF radio every four hours, giving the position, speed and direction of travel of Ignacio. We plotted the track of the hurricane as it approached the Big Island. Finally, after two days, the hurricane veered off to the east and diminished in strength. It would miss the Big Island. At 5 pm that day the hurricane watch was discontinued. It was a great relief. Now we could all relax and enjoy our stay in Radio Bay.

## Sailing to the Kona Coast

The day after the hurricane passed by Dan's folks flew from Oahu to the Kona Coast of the Big Island, where they would stay at the luxurious Kona Coast Resort for a week. To visit them we would have to sail around the northern end of the Big Island. We left Hilo in late afternoon, and sailed all night, standing three-hour watches.

This was easy sailing. The waters were fairly smooth, the breezes warm, and we were always in sight of land. The next afternoon we entered a beautiful small cove on the Kona side of the island adjacent to the resort. It was a surprise to find two other sailboats anchored in this small cove. Fred and Jo from San Francisco were on board their boat, a CT 41. Fred was a net control on the Pacific Maritime ham radio net for many years. (The net control is in charge of the net, conducts the daily roll call, and handles any emergencies for vessels underway.) They were experienced sailors and were now working on Oahu. Jack Ayres, on *Compadre*, a Tartan 34, was an oceanographer who had been cruising for more than 10 years. He had written a book called *Sea Creatures*, which we happened to have on board. It was delightful to meet him and have him autograph our copy. Before we left Jack snorkeled over to the area where we were anchored. He noticed that our anchor chain had wrapped around two coral heads due to the boat swinging around with the wind and currents. It was the first time we had anchored near coral, and we didn't realize this could happen. Jack directed us as to which way to turn as we motored slowly around to un-wrap the chain from the coral, until we had it straightened out. We were grateful for Jack's help. It would have been very difficult to disentangle the chain ourselves.

Every morning we went ashore to join Dan's family for smorgasbord breakfasts and lunches at the resort. We took tours of the gardens with many beautiful tropical plants, and had delicious ice cream cones and wonderful dinners. Then we took nice hot showers before returning to the boat to sleep. We talked all about our passage, omitting some of the scary parts so they wouldn't worry about us or question our sanity for sailing to Hawaii and back. Dan's folks had never been to Hawaii, and were enjoying it immensely, as were we. We had stopped in Hawaii on our flights to and from Australia for short vacations, but this was the best Hawaiian vacation we'd ever had. It was wonderful to be afloat in these beautiful warm waters, surrounded by tropical vegetation and be able to visit with Dan's family at the resort.

After five days Dan's family flew home, and we began readying the boat for our passage back to Seattle. Dan installed our new radio, and we bought a bottle of rum to help me sleep on my off-watches. We needed to be back for the start of school, the day after Labor Day, about a month away. The days were warm and breezy, and we were looking forward to a much better passage than the one over to Hawaii.

## Trip Back from Hawaii

We left the Kona Coast of the Big Island on August $4^{th}$ in strong winds and big seas. However, the skies were clear and the breezes warm. The sailing was much more pleasant than when we left Neah Bay for Hawaii in June during a gale. We saw a pod of dolphins as we headed northeast to clear the island of Maui. Our new radio worked well. Every day we checked into the Pacific Maritime Net to give our position and our weather conditions. It was comforting to know people were keeping track of us and could give us a forecast of the weather in our area. Dan also had conversations with his father twice a week using Morse code. About a week out we were lucky to spot a green glass Japanese fishing float about a foot in diameter. We sailed over to it, but it took three tries to pick it up. It was too large to fit in our homemade fish net. Dan nearly fell overboard on our second attempt. He didn't have his harness on. On our third attempt Dan started the engine. When we came along side the glass float I used the net to hold it against the hull, then reached down and grabbed the attached rope and hauled it aboard. After Dan scrubbed off the goose barnacles we had a beautiful souvenir of our passage home.

We made good progress and by the 11th day we calculated we were at the halfway point—1,300 miles from Hawaii, and 1,300 miles from Neah Bay. If we continued at this speed we should be there in 22 days, a week before the start of school. Unfortunately, the next day we sailed into the Pacific High. This is a high-pressure system that settles over this part of the Pacific in summer months. Normally we would be able to stay in the wind by sailing over the top of the high. This summer the Pacific High was so large that it was impossible to sail over the top of it. We couldn't motor through it either, because our boat only gets 10 miles per gallon of diesel, and we had less than 20 gallons of diesel left in the tank. There was nothing we could do but wait for the wind to pick up. For the next 10 days we had light, fluky winds or complete calms. Although we took advantage of every puff of wind, we sailed aimlessly, making little progress towards home. Every day Dan became more and more frustrated. It looked as if we would be late for the start of school.

To me this calm spell was pleasant—much preferable to the strong winds and big seas we had on the way over to Hawaii. There was a gentle rise and fall of the boat as we floated on the undulating surface of the sea. It was easy to stretch out on the foredeck and read, or gaze at all the sea life in the deep clear blue water. We saw two small whales in the distance dive repeatedly. As they approached our boat they dove deep and disappeared. They didn't appear menacing, and we were disappointed we didn't get to see them up close. One day we had three albatrosses paddling around behind our boat like giant ducks. We had seen lots of them soaring over the waves, but never up close like this. Like us they were stranded, with not enough wind to fly into that would lift them off the water. Blue bottles, also called by-the-wind-sailor jellyfish, were gathered by the thousands in the calm waters. It was like a sea of jellyfish. These creatures have transparent floats from which long bright blue tentacles are suspended. They move along the surface of the water, propelled by the wind. They also were unable to move until the wind returned. In this calm weather I was able to bake bread, dinner rolls, casseroles, cakes, and brownies. It was easy to sleep, although we still had to keep three-hour watches around the clock. We saw many freighters, and were able to talk to the officers on the VHF radio and get a position report. It was pleasing that the positions they gave us agreed with the positions we calculated using the sextant. It gave us more confidence that we could navigate with a sextant.

When the wind picked up again we began catching tuna—five altogether in the course of a week. Most of the tuna were only 3 to 4 pounds. The last one was 23 pounds, 36 inches. We had delicious fish lunches and dinners, and pickled the remaining fish in vinegar and spices to eat later.

One evening I planned to make a chicken noodle casserole and some chocolate chip brownies. After I mixed the ingredients I tried to light the kerosene oven, by squirting a small amount of alcohol primer in the cup below the burner. Apparently, kerosene had leaked from the burner earlier and filled up the pan underneath the oven to a half-inch deep. When I lit the alcohol primer the kerosene in the pan caught fire, sending up bright yellow flames. The flames quickly built up, rising to two feet above the stove almost to deck level. I had visions of the fire spreading and burning the boat to the waterline. I shouted "Fire!", but Dan was unperturbed. He came down the ladder from the cockpit, grabbed our small Halon fire extinguisher and with a push of the button sprayed the flames beneath the stove, putting out the fire in a few seconds, all the while listening to music on his Sony Walkman. I couldn't believe his nonchalance! The Halon fumes inside the boat were terrible. I had to go out on the foredeck and wait for a half hour for the fumes to disappear so I could re-light the oven and cook our dinner.

On the 21st day we finally began to get the westerly winds that would push us the rest of the way to Neah Bay. We were so happy to be sailing again that we put up the maximum sail area to take advantage of the following wind—the full main on one side and poled-out genoa (our large foresail) on the other side. At last we were moving along nicely in the right direction. The winds gradually increased during the evening. During my watch from midnight to 3 am I watched our speedometer increase to eight knots. When I looked around at 2:30 am I was alarmed to see that the boat was way off course. The self-steering was no longer working because the push rod had popped out of position in the huge seas. The boat had rounded up (headed up into the wind). In the strong wind and heavy seas it would take a lot of strength to force the tiller over to get back onto our downwind course. I woke Dan, and he quickly got up, climbed into the cockpit, and pushed the tiller over. The wind caught the mainsail, forcing it over to the other side. This sudden jibe caused the webbing of the boom vang to break. (The boom vang fastens the boom to the deck and prevents the boom from swinging across when the boat jibes). It was our first gear failure. It was pitch dark, with heavy rain and gale force winds. By this time the seas were enormous. It was too dangerous for Dan to go out on the bowsprit to take down the genoa, or to hang off the stern of the boat to try to fix the self-steering mechanism. Instead, he took down the mainsail, leaving the genoa poled out. This reduced our speed somewhat, but we had to steer the boat by hand. In the darkness I found I was unable to keep the boat on course, because I rely on the telltales, streamers of yarn tied to the shrouds, to determine the direction of the wind. They are visible from the cockpit during the day, but were not visible in the dark. Dan had to hand-steer the boat for the next three hours, while I went below to try to sleep.

Finally, at dawn, I took over the steering, so that Dan could take down the pole and the genoa, raise the main, put a reef in it, and repair the self-steering vane in daylight. It was a great relief to be sailing again with the boat under control with the right amount of sail.

We continued on this course for five days, then in the early morning of the 27th day Dan sighted the tall peaks of Vancouver Island lit by the rising sun. We had made landfall! We were only 120 miles from Neah Bay. As we approached the entrance to the Strait of Juan de Fuca at 1 am the next morning a thick fog suddenly descended. We had seen two tugs behind us, but now in the dense fog we could only see the lights on the superstructure of the tugs and of ships in the area and hear the repeated deep blast of their foghorns. We were in the middle of the shipping lanes, with ships all around us, and we had less than 50 feet of visibility. The wind had died and Dan was on watch motoring at full speed (five knots), stopping every few minutes to listen for engines of passing ships. This was the scariest time of the whole trip.

I went below to sleep, but at one point I felt our boat veer suddenly to the right. A freighter had appeared in front of us with its two range lights aligned vertically (this meant the ship was headed straight for us). Dan had had to push the tiller hard over to avoid a head-on collision. After five hours I took over the steering so Dan could go below and get some sleep. As I motored through the fog a fishing boat suddenly appeared out of nowhere. I had to quickly alter course to avoid a collision. As we got closer I could see the fishermen on the aft deck of their boat tending their lines. When the fishermen looked up they appeared to be very surprised to see a sailboat in their midst. This happened 10 or 12 times during the next half hour. Looking at the chart later we realized that we had unwittingly passed over the Swiftsure Bank, where dozens of fishing boats were trolling for salmon. The fog didn't lift until 10:30 am, revealing the tree-covered hills of the Washington Coast. Early in the afternoon of Labor Day we finally reached Neah Bay, where we could stay overnight and get some sleep before heading down the Strait of Juan de Fuca and south to Seattle. We were exhausted, having been awake for most of the last 24 hours. We pulled into the dock to fill up our water and diesel tanks, and Dan found a pay phone so he could call his family to tell them we had arrived. After dinner we fell asleep immediately.

The next morning at 8 am we left for Port Townsend. There was no wind, and the Strait of Juan de Fuca was the glassiest we had ever seen it. We motored for 14 hours, arriving at the marina in Port Townsend at 10 pm. Dan was a day late for his teaching job, and it would be another day before we could make it back to Seattle. When we got up the next morning we bought a newspaper before departing. The headlines announced "Seattle Teachers Out on Strike". For Dan this was a tremendous relief. Instead of starting teaching two days late he could relax. During the strike he would have time to work on the varnish, which had deteriorated badly in the sun and saltwater during our trip. We took a leisurely trip back to Seattle that afternoon. Our ordeal was over. I told Dan I never wanted to make another passage.

# 3 CALIFORNIA AND MEXICO

## Planning a South Pacific Cruise

After our shakedown cruise to Hawaii we settled back into our life of working from September to June, and sailing locally in the summer. The first summer we took our usual cruise to Desolation Sound, enjoying the light winds and calm seas. The next summer Dan brought up the subject of another offshore cruise, this time to the Queen Charlotte Islands, off the coast of British Columbia. We would sail up the Inland Passage as far as Prince Rupert, and then sail out to the islands.

Visiting these islands has much in common with visiting islands across the Pacific. They are remote with very few visitors, although they are only a one-day sail from the British Columbia coast, instead of a passage of several weeks. Often we would find we were all alone in an anchorage. We had to be self sufficient, as there are no marinas or repair facilities, and few grocery stores. We had to carry spare parts for the boat equipment and engine, and be able to repair anything that failed. We found these remote islands fascinating, with their ancient Indian villages, and their mining, logging, and whaling history. Sailing to the Queen Charlottes gave us a taste of what it would be like to sail to the islands in the South Seas.

The trip was uneventful until our return to the mainland in late August. After crossing the Strait of Georgia headed to Newcastle Island in Nanaimo we were unable to start the engine. After many attempts with no luck, we decided to sail the short distance into the anchorage at Newcastle Island. As we approached the anchorage we had to drop the mainsail to slow down, sailing slowly with only the small staysail. Off our starboard side we saw a runabout about a half mile away motoring while the skipper faced the stern of the boat, working on the outboard motor with the cover removed. We could see that if he continued on his course his runabout would collide with our boat. With our engine not working and only a small amount of sail up we were unable to maneuver out of his way. Dan and I shouted as he got closer to warn him we were in his path, but he didn't hear us over the roar of his outboard. He never looked forward until with a horrible crunch his boat hit our sailboat square in the middle of the starboard side. The impact threw him sprawling forward into the bottom of his boat and separated the hull from the deck of his boat. Fortunately he wasn't injured. He claimed that when he had started out across the bay the way was clear, but he obviously should have looked up from his outboard as he motored along. At first we saw no damage to our boat. On closer examination we realized the bow of his boat had hit the strongest part of our hull, the wide, quarter-inch-thick stainless steel chain plate, attached at an angle between the rub strake and the hull. The force of the impact put a permanent bend in the thick stainless steel, but otherwise did no harm. The skipper offered to pay for any damage, but we didn't think we needed any repair.

Over the next few days we tried to have the engine worked on, but the only Volvo diesel mechanic in Nanaimo was too busy to help. By the greatest luck Betty and Tony, friends of ours, were at anchor at Newcastle Island on their newly acquired Grand Banks powerboat. They had witnessed the collision, and were sympathetic with our plight. They offered to tow our boat back to Seattle so that we could return in time for Dan's teaching job, and we could have the engine repaired near home. This was a godsend. Betty and Tony were experienced sailors, and for three days they towed us through Dodd's Narrows, to Friday Harbor for customs, across the Strait of Juan de Fuca, and all the way to our slip at Shilshole. While being towed we had to stay on our boat to steer and keep the boat from slewing around. During the days we were towed it felt like we were on a magic carpet. It was heavenly to be travelling through the water at six knots without any sound except the distant hum of the engine on the powerboat pulling our boat on a long towline. Every evening they towed us into a harbor where we could drop our anchor, and they anchored near by. After we were anchored, they had us over for a delicious dinner that Betty had prepared while underway.

After a couple of months getting estimates in Seattle for repairing the engine, which turned out to be about $4,000, we found out that the builder of our boat could buy a new Yanmar engine for us for only $5,000 so we chose that route. With the help of Al Hughes, a mechanic friend at Shilshole, Dan installed the new engine over the winter months, and by June the boat was ready to go.

After we returned from the Queen Charlotte trip Dan again brought up the subject of another offshore cruise. Dan had enjoyed the challenge of sailing across the ocean, and thought about making another passage soon after we returned from Hawaii. He proposed taking a year's leave of absence from his teaching job, hopefully extending it to two years. We could leave in August, sail down the West Coast to Mexico, then depart for French Polynesia in April or May. This would mean we could avoid the hurricane season in Mexico (June to November), and be in French Polynesia for six months during the non-hurricane season in the Southern Hemisphere (May to October), then spend the winter and spring in Hawaii before returning to Seattle in the summer. This would be a fairly leisurely trip, allowing much more time for visiting islands than making passages. My vivid memories of the many difficulties of our Hawaii trip had faded somewhat. Dan's enthusiasm for ocean cruising and the lure of the South Pacific islands finally convinced me to try ocean sailing again.

The only part of the proposal I didn't like was sailing down the coast from Neah Bay to San Francisco. The worst part of our cruise to

Hawaii had been the cold, rough sail down the Washington and Oregon coast in gale-force winds. I wanted to avoid that if at all possible. Instead, I proposed taking Amtrak to San Francisco, and meeting the boat in Sausalito for further cruising. Dan asked a couple of friends if they would like to crew down the coast, and they readily accepted. Both were boaters in local waters and wanted to get some offshore experience. It was a great relief for me to avoid that most difficult sail off the coast of Washington and Oregon.

Although Dan was able to get a leave without pay from his teaching job, I had to quit my job as a research tech at University of Washington, and take my chances at finding another job when we returned. Having worked in Seattle for 14 years while living a frugal lifestyle, we had saved enough money to afford a two-year cruise if we kept our spending down to $1,000 a month. Since we would not be paying rent, moorage, utilities, car insurance, or car repair, we figured our expenses would be much less than living on land.

After Dan's teaching job ended in June and I quit my job, our days were consumed with preparations for the trip. For our trip to Hawaii we had equipped our boat with a self-steering windvane, an auto helm, and a life raft. We had also had a small dodger installed. This was a frame of curved stainless steel covered in fabric mounted over the main hatch to provide protection from the weather. Now we added solar panels to recharge the batteries, and a Sat Nav to replace the sextant for navigation. I sewed a new, larger awning for the cockpit to keep the tropical sun and rain off us when at anchor, and a new Sunbrella cover for the wooden fore hatch to protect it from sun and saltwater.

During the weeks of preparation I began to have doubts about the wisdom of making this two-year voyage. Could I overcome my seasickness, which was induced by ocean swell? Could I overcome my fear of whales? Could Dan and I get along in the small space of a 28-foot boat for two years? Once the decision was made, we were committed to this cruise. There was no turning back—no jobs, no house, and no car waited for us in Seattle. It was scary to contemplate being at sea for two years, away from the comforts of home, but it was also exciting to be starting on a new adventure. After much last-minute preparation and provisioning the boat was ready to leave Seattle on our planned departure date of August 15th.

## Harbor-Hopping Down the Coast

On August 15, 1988 a small crowd gathered on the dock at Shilshole to wish Dan and his crew, Tom and Ethan, "Bon Voyage". They were departing for a one-week trip to San Francisco. The date was set months before and amazingly they were able to make it, considering all the work it took before leaving. Not only did we have to get the boat ready, but in the weeks before leaving we sold or gave away most of our furniture and belongings, packed up items we wanted to keep and put them in storage, moved out of our rental house, sold our car, and bought provisions for the boat. As I said goodbye to Dan and crew, I was sad to see our home sail away, but glad it was in such capable hands. For now I would be staying at my sister and brother-in-law's home in Wedgewood until I could re-join Dan and the boat in the Bay area.

My trip down the coast on Amtrak was easy. Dan's cousins, Norm and Judy, picked me up from the Sacramento train station and drove me to their home in nearby Roseville. They were wonderful hosts, and we enjoyed a few days' visit before the boat arrived at the dock in Sausalito. Eight days after their departure, Dan and his crew arrived. They were in good spirits, and the boat was in great shape. Dan explained why their trip took longer than expected. They were hit by a gale off the northern California coast with winds of 40 knots and huge seas. They had chosen to heave-to for three days, essentially parking the boat, afraid that if they continued sailing downwind waves would break into the cockpit, possibly soaking everything down below. My eyes got big when Ethan described the 12-foot waves they experienced. If I had been aboard I would have been so frightened I would probably never make another passage. Dan had enjoyed the passage. He regarded the gale as just one of the challenges of sailing. He had turned out to be much more of a sailor than I.

Our cruising in San Francisco Bay was a whirlwind of activity. It was easy sailing from one place to another. For a few days we moored at the city marina where we took the bus to Golden Gate Park to visit the zoo, aquarium, and planetarium. From there we took a cable car through Russian Hill and in the evening visited the Exploratorium, the best hands-on science museum we had ever visited. Best of all, it was free. We sailed to Berkeley Harbor Marina to visit Phil, a fellow graduate student, now a U.C. professor. Dan's cousin from San Carlos, Linda, came down with her daughter Melinda and two children to see our boat when we moored at Redwood City Marina. Sailing friends Ken and Carol came down to take us to their home in Los Altos for dinner, and we took them for a day sail on the bay the next afternoon. In mid-September we sailed up the channels of the Sacramento Delta, through many opening bridges, anchoring at Steamboat Slough, Potato Slough, and Georgiana Slough. For three days we moored at Four Seasons Marina in West Sacramento to visit Dan's cousins Judy and Norm. They took us sightseeing and to their home. While at the marina we watched a fisherman haul in the largest fish we had ever seen—a seven-foot sturgeon, weighing over 200 pounds. On our last day in the Delta our depth sounder quit working. It was a harrowing experience casting out a lead line to determine the depth as we crept along to an anchorage, and the next day as we motored out to the exit of the Delta into the deeper San Francisco Bay. We would have to take the depth sounder in for repair before we left the Bay area. It was the third instrument to fail on this trip. Our auto helm and Sat Nav were already in the shop being repaired.

Along the way we met John and Wendy on *Alobar*, Jan and Tom on

*Ambler*, and Jan, Dave, and son Joel on *Moulin Rouge*, shared dinners with them and discussed our plans for sailing to Mexico and French Polynesia. Later we met Jim and Norrie, from *Allele*, another Bristol Channel Cutter. They had been in the same gale off the northern California coast, but had continued running downwind. A huge sea broke over their stern, heeling the boat over 120 degrees. Jim was underwater, hanging onto the boom gallows till the boat righted itself. Norrie was in the cabin down below, but opened the hatch to see if Jim was all right. Tons of water came through the open hatch, drenching the inside of the boat. When they reached port Norrie didn't even want to talk about the passage. It took 16 loads of laundry to get all the salt out of their clothing and bedding. A Canadian man we met had sailed down the coast with his teenaged daughter and her boyfriend. All of them were seasick the whole way. After that passage he cancelled his plans to take a one-year voyage to New Zealand, and instead decided to ship the boat back to Canada.

At the beginning of October we began our sail down the coast to San Diego. This is far different from sailing down the coast from Seattle to San Francisco. Although it's the Pacific Ocean the light winds and slight seas made it much more enjoyable sailing. We sailed short distances between anchorages, mostly in daylight. The only difficulty we encountered was fog, which usually lifted by noon.

Monterey Harbor was one of our favorite places. The aquarium is the best we've seen, with its 30-foot-tall tank containing a kelp forest teeming with sea life, including sea otters drifting around on their backs on the surface. There was a small aquarium full of snapping crabs. We had been puzzled for years by the crackling noises we heard on our boat at many of the anchorages in Australia and California. We learned that the noise is made by tiny crabs that attach themselves to the hull, and snap their small claws together with surprisingly loud clicking sounds.

Our next anchorage was Stillwater Bay, where we tucked into a small cove after threading our way through masses of kelp. It was wonderfully quiet, after listening to the sea lions in Monterey Bay barking throughout the day and night. It was entertaining to watch the sea otters swim around amidst the kelp, diving and coming up to crack clams on their chests as they floated on the surface. The next day we watched a large commercial boat sweeping across the bay harvesting kelp which fed into their open stern. It was like a giant floating vacuum cleaner.

At San Simeon we anchored off a beach adjacent to Hearst Castle and rowed ashore to take a Parks Department tour of the property. Although we got soaked to our knees when our dinghy was hit by a breaking wave on the beach, we enjoyed the tour, squishing through the premises in our squeaking wet TEVA sandals. We rode the bus up the hill to the estate, our ears popping the whole time, having been at sea level for so long. The extravagance of the estate with much furniture from European churches, a huge indoor swimming pool, and a long table for 20 guests was in startling contrast to our small boat, with its tiny floor space, barely room for two people, and no table at all.

The next anchorage down the coast was Morro Bay. The Morro Bay Yacht clubhouse is located by the wharf, and the members were very friendly to visiting cruisers. Every Wednesday night, cruisers from the boats tied up to the wharf joined the yacht club members for a dinner of hamburgers, which you grill on their barbecue, with toasted buns, beans, salad, and ice cream provided by yacht club members. At a cost of only $2.75 per person it was real bargain. We had a delightful time, staying till nearly midnight chatting with other cruisers.

San Miguel Island, in the Channel Islands, was one of the last anchorages before San Diego. As it was my birthday, I signed up for a free tour of the island given by the ranger. Dan preferred to stay on the boat. On the VHF radio I arranged to meet the ranger at the small ranger station southwest of the dinghy beach the next morning.

However, when I got ashore I wasn't sure what direction was south or west, and I ended up trekking from one side of the island to the other for three hours before I spotted the small rangers' headquarters. The ranger was surprised it had taken me so long. He told me that in the year he'd been a ranger no one had ever got lost. This didn't bode well for my being able to find the way to the Marquesas. The ranger was happy to take me on a tour to places I hadn't seen on my trek, and explain some of the history of the island. The island had been grazed by sheep for over a century, and was now being restored to its natural state. He showed me fox trails, sea birds, and a large colony of seals and sea lions. It was a wonderful birthday treat.

It was great to visit the California coast by boat, and meet other cruisers starting out on their cruise to Mexico and French Polynesia. Harbor hopping is my favorite kind of cruising, and it was an excellent beginning for our two-year cruise.

## Bluegrass Music

We met Cris and Dick in San Diego in November 1988 about a week before we left for our trip to Mexico. We saw their boat, *Chatauqua*, at the San Diego Yacht Club. It was a Bristol Channel Cutter, a sister ship to our boat, except it was more traditional, with a wooden cabin top, a gaff-rigged mainsail, and wooden spars. They invited us aboard to see their interior. We were very impressed. They had finished the boat themselves, starting with a fiberglass hull made by Sam Morse. It was beautifully finished, with stained glass windows on the forward cabinet doors, and three lockers in the main cabin designed to hold their musical instruments—banjo, guitar, and mandolin. On our boat that space was used for food storage, and I had no special cabinet for my violin. I stuffed it into the quarter berth wedged between sleeping bags and blankets to protect it from wild motion when we sailed in rough seas. Cris and Dick were leaving for Mexico in a few days, so they and we were very busy buying last-minute equipment and provisions before departing. We hoped to meet again when we reached Mexico.

A week later we left for Mexico, stopping at Cedros Island and Turtle Bay on the way down the Baja coast. The crystal clear water and the abundant sea life amazed us. We watched pelicans, seals and seagulls all diving at once in a feeding frenzy. There were sea lions and elephant seals on the beaches. As soon as we were anchored fishermen came up to our boat in their pangas (long open fiberglass fishing boats powered by an outboard engine) to offer us lobsters or shrimp either for trade or money. Lobsters were only a dollar each, so we feasted on them whenever we had the chance.

Several weeks later, when we sailed to Bahia Santa Maria, an anchorage on the Baja coast, we met Cris and Dick again. They came in, anchored near our boat, and invited me over with my violin for a jam session. They had been playing and singing bluegrass in California, and had a large repertoire of bluegrass, blues, folksongs, and sailing songs. One of our favorites was a John Denver song, "Blow up your TV". Dick played both banjo and guitar, Cris played mandolin, and they sang in beautiful harmony. They had left their fiddle player on land (he was not a sailor), and wanted me to join them to form a bluegrass band. I loved to hear Cris and Dick sing and play, and tried to play along when I could. It was a real challenge, as I had only played classical music, couldn't play without music, and didn't know any of the songs. Cris loaned me her mandolin book of bluegrass tunes, and I learned to play six or eight tunes, such as "Blackberry Blossom", "Whiskey before Breakfast", "Red-haired Boy", and "Soldier's Joy". All through December we had frequent jam sessions either in the cockpit of their boat or on our boat. Dan joined us to listen and offer encouragement, and we all had happy hour afterwards. Sometimes friends would join us to play or sing along or just listen to the music, and we were invited to play on other people's boats. At the next anchorage, Magdalena Bay, we resumed our bluegrass sessions. Dan said we were sounding better and better.

We left several days later to sail to Cabo San Lucas, at the southern tip of the Baja Peninsula. We had a very rough sail, leaving in late afternoon and fighting strong winds the whole time. It wasn't till the next evening that I was able to cook any food—a simple dinner of two cans of ravioli heated up in a pot. I ate mine while Dan was on watch. Unfortunately, by the time he came down to have his it was burnt. We were both exhausted from lack of sleep.

*Chautauqua* left right behind us and had a harder time than we did. They had to hand steer because it was too windy to put up a headsail, and their self-steering didn't work with the main alone. On the morning of the third day Dan woke me at 5am for my watch and told me the sun was just rising, and he had seen *Chautauqua*'s masthead light behind us. Dan must have been really groggy. It was beginning to get light, but the sun didn't rise until 6:30. *Chautauqua* was reported to be 15 miles behind us the previous night, and couldn't have made up that distance overnight. We had had to reduce sail to slow down to avoid arriving in the dark. We finally made it into Cabo San Lucas in mid-morning, made a quick trip ashore to check in and buy fresh food, and then napped all afternoon. *Chautauqua* arrived in the afternoon during our naps. We woke up to have a quick dinner and went right back to bed.

The next day we got together with Cris and Dick for a jam session and rum drinks on *Chautauqua*. Cris said they were celebrating just being alive after their horrible passage from Magdalena Bay. While in Cabo I was able to find a photocopy shop, and copy pages from the mandolin book of bluegrass tunes.

From Cabo San Lucas we went our separate ways. Dan and I sailed to Isla Isabella and down the mainland coast, eventually ending up at Puerto Vallarta, the jumping-off port for French Polynesia. Once again we got together with Dick and Cris for bluegrass sessions, taking up where we had left off. On St. Patrick's Day the cruisers got together for a party at the marina. It was drinks and hors d'oeuvres on the docks. We were invited to play bluegrass while strolling up and down the docks. I still needed to read the music, so Dan agreed to be my music stand. I just pinned each piece of music onto the back of Dan's T-shirt with clothespins, and he preceded me as we strolled along. It was difficult to keep pace with Dan and also pay attention to my footing. After that experience I decided to memorize the music so in the future I could play bluegrass music whatever the conditions—strolling along the dock or playing on shore in the dark.

## Whale Watching

When we made our first offshore trip to Hawaii I had two fears: fear of seasickness and a fear of whales. Every summer when we crossed the Strait of Juan de Fuca I suffered from queasiness if there was any ocean swell. If I took Dramamine I was so sleepy that I couldn't enjoy the sail. My fear of whales was fueled by stories we had read of sailboats being sunk by whales while crossing the ocean. The only whales we had seen in the Pacific Northwest were Orcas, which we were always delighted to see. My fear was of the giant whales in the Pacific Ocean, such as gray whales and humpback whales.

We solved the seasickness problem on the way to Hawaii by using scopolamine patches behind our ears. They had to be replaced every three days. They were completely effective in preventing seasickness, but the side effects were unpleasant—food tasted strange, we had dry mouths, and our vision was blurred, making it difficult to read the charts.

Three years later, when we sailed down the West Coast to Mexico, we had a cure for my seasickness. A cruising couple we met in Canada the summer before told us about meclizine, the generic name for Bonine. A bottle of 100 pills cost about $4, and a pill lasted for 24 hours. For me it prevented seasickness completely, and had no side effects. Now my main fear was of whales.

During the 24 days to Hawaii and 28 days returning to Seattle we saw only one pair of whales that appeared in the far distance. As they approached our boat 100 feet away they both dove down deep. They never came up to the surface again so we could get a closer look and identify them. When I realized how few whales there were out in the Pacific my fear of whales began to fade.

On our way down the coast of Baja, Mexico we stopped at Magdalena Bay. Shortly after we arrived Dick and Cris on *Chautauqua* came in and anchored near by. On a trip six years previously they had spent many weeks in Magdalena Bay with friends on another boat. They had the exciting experience of watching the gray whales during their migration south come into the lagoon to give birth. Now it was December, time again for the migration of the gray whales. We were hoping to see the whales enter the bay anytime. Cris, Dick and we stayed at Magdalena Bay for over two weeks while waiting for the whales. There was plenty to do. We feasted on a 20-inch tuna we caught on the way to the bay. We traded T-shirts for lobsters from Mexicans, bought bowls of shrimp for only $3, and scooped up buckets of scallops from the beach just below the low-tide line. Dick and Cris were great company—we hiked across the dunes collecting shells, sailed our dinghies in the flat waters of the bay, and swam in the clear water where we were anchored. We celebrated Christmas on *Chautauqua* with a shared dinner of baked canned ham with cloves and pineapple, baked beans, cabbage slaw, and chocolate chip brownies. Still no whales appeared.

The day after Christmas we headed south to Cabo San Lucas. Cabo was a gathering place for cruisers, with 30 or more boats at anchor in the open bay. After a week at anchor at Cabo I saw my first whale. It spouted three times near the beach. The migration had begun!

In the afternoons we congregated with other cruisers at Hotel Finistere for $1 margaritas and whale watching. We sipped our margaritas on the patio, where we had a good view over the beach out to the ocean. The gray whales came in close to shore and we watched them spout and occasionally breach. Then one afternoon a pod of whales came into the anchorage where two dozen or more sailboats were anchored. We knew when a whale was in sight when we heard nearby cruisers screaming in delight. Everyone ran out to the bows of their boats to see the spectacle. It was amazing watching these giant creatures surface in the maze of anchored boats and send a tall plume of mist gushing from their blowholes with a loud whooshing sound. People pay lots of money to go out and search for whales on whale-watching trips. Here we had a free whale-watching show in our backyard. It kept us all entertained for hours, as one after another whale rose to the surface and spouted between the boats. In spite of their size, more than twice the size of any of the boats, there was no fear that a whale would come up under a boat and capsize it. The whales seemed perfectly capable of finding clear water in which to come up to breathe.

This experience cured me of my fear of whales. The incidents we read about of boats colliding with whales were rare. Now that I was familiar with giant whales they didn't appear to be menacing. I welcomed the sight of whales and hoped we would see more of them as we crossed the ocean.

## Dolphins

One of the delights of cruising is watching dolphins play around the bow of the boat while sailing. Usually we hear the repeated high-pitched squeaks of dolphins when we're down below in the cabin. The sound carries through the water and through the fiberglass hull, sounding like a child's toy that squeaks when it is squeezed. Sometimes we see a pod of dolphins from a distance and watch them as they approach our boat, dash ahead to the bow, gracefully arch out of the water, and crisscross in front of the bow. They repeat this over and over, taking turns coming up on either side of the boat. We're always thrilled when they join our boat. It's not known why they play at the bows of boats, although some people think they enjoy the boost in speed from the bow wave that is formed in front, much as we like to body surf in waves on the beach. Sometimes they stay with our boat for 10 or 15 minutes before disappearing into the distance. We always wish they would stay longer.

When we sailed down the California coast we were fascinated by phosphorescence in the water. Tiny sparkles appeared as we pumped saltwater through the bowl of the toilet. At night, when we returned from visiting other cruisers at an anchorage, we could see the outline of the oars as we rowed back to the boat. When we raised the oars out of the water the drops of falling water produced bright concentric rings. It was fun as a passenger in the dinghy to drag a hand in the water and see the beautiful sparkles forming in its wake. On deck, we enjoyed scooping a bucket of saltwater and pouring it out over the side of the boat just to see the display of beautiful sparkles wherever the water is disturbed.

On a dark night, as we sailed along the coast of Southern California, we noticed streams of phosphorescence in our wake and in our bow waves. As our bow cut through the water short streaks radiated out from the boat, indicating fish darting away from our path. In the middle of the night we heard the squeaks of dolphins in the water around us. We were fascinated by their bright outlines as they leapt out of the water and plunged back in along the side of our boat. From the phosphorescent trail they left behind we could see how dolphins moved through the water. They swam past our bow, made a large circle, and ended up behind the boat. They then turned around in our wake and came up alongside again and up past our bow. They were swimming in wide circles at a much faster speed than we could travel. It was a magnificent show.

We met a cruiser along the coast who told us he'd had dolphins playing at his bow for an hour or more. He has a large dog on board that gets excited when dolphins appear, running up to the bow and barking loudly. The cruiser thinks the dolphins respond to his barking dog and are entertained by its presence. The next time we saw dolphins we decided to test this theory. We took turns standing on the bow of the boat and barking loudly like a dog. We felt foolish, but there were no other cruisers in the vicinity to hear us. The dolphins continued to cavort at our bow for a while, but we were disappointed when they disappeared after the usual 10 minutes or so. We suspect that we didn't sound much like real dogs. Dolphins are fairly intelligent animals, and probably not easily fooled.

When sailing in the ocean we enjoy seeing seals, skates, rays, and whales, but dolphins are the most entertaining. We're entranced by their smooth, graceful motion while swimming and their exuberance as they leap out of the water. They're a pleasure to watch because they seem to be leaping for joy.

## Las Hadas

After a couple of months cruising down the Baja coast, we headed for the Mexican mainland. We had several months to explore the many anchorages on the mainland before we planned to leave for French Polynesia in April. One of our favorite anchorages was Las Hadas, south of Puerto Vallarta. The first week in March 1989, we sailed to Las Hadas, where we anchored in the harbor in front of the huge Las Hadas resort hotel. The picturesque hotel is built on a cliff overlooking the sea, with multi-layered rows of rooms all connected in hodge-podge fashion. Variously shaped windows looking out to the sea are cut in the white concrete walls and the units are topped with red tile roofs. It appeared to be somewhat rundown, and was almost completely vacant. The hotel staff allowed cruisers to come in and swim in their beautiful swimming pool for free. It was wonderfully refreshing in the hot afternoons.

The day after we arrived we were able to check in at the nearest port of Manzanillo, about five miles south in Manzanillo Bay. At that time cruisers were required to check in on arrival at every major port they visited. Checking in usually meant taking long walks in the hot sun to customs, the Port Captain, and immigration offices located in three different buildings in the small villages. There were four other boats at Las Hadas when we arrived, and we decided we would all go in one boat to save each boat making the one-hour trip to Manzanillo separately. Bob and Marge offered to take us and the crew of the three other boats over in their 35-foot classic wooden boat *Tusitala*. The next morning we rowed over to *Tusitala* along with six others (Eddie and Eileen from their boat *Nubian*, Bob and Michelle on *Akavit* and Dorothy and Bill on *Zingaro*) for the trip to Manzanillo. So that we could go ashore when we arrived three dinghies were towed behind *Tusitala*, ours and two inflatables. After we anchored at the large port of Manzanillo we took Bob and Marge ashore in our dinghy. Our dinghy is a fiberglass sailing dinghy, which has a three-foot long vertical centerboard for sailing. When rowing, we use a short plug in the centerboard trunk to keep water from coming up through the space at the top. Unfortunately, when all four of us were in the dinghy the top of the centerboard trunk was just at sea level. As Dan rowed the seawater seeped through the centerboard trunk, where Dan was sitting. Bob and Marge were seated in the stern, which was dry. As Dan and I sponged the saltwater out we tried to act nonchalant so Bob and Marge wouldn't think the dinghy was about to sink. The dinghy was fine with two or three people, but we realized that our dinghy doesn't hold four large adults. It was only a short distance to the ramp in back of Rey Coleman's restaurant, so we made it without shipping too much water. Dan forgot his shoes, but was able to borrow a pair of work boots from Bob—jumbo size.

It was an easy check-in procedure, with the Port Captain and immigration located in the same building. We five couples walked a few blocks to the Port Captain's office with our passports, boat documentation and crew lists and paid a very small port fee ($1.17), but we also contributed to the Red Cross when the immigration officer held out a cup. On the way back Dan and I stopped at the Conasupa, part of a chain of inexpensive Mexican markets where most of the shoppers are Mexican. We bought as much food as we could carry, including brown rice, guava jam, and lots of fruits and vegetables.

By the time we finished the check-in procedure it was time for a late lunch. We all flocked to Rey Coleman's seafood restaurant. Rey's claim to fame is that he went over Niagara Falls in a barrel. We had seafood soup, which is an unusual dish in Mexico—it was delicious. By the time we finished lunch it was almost 4 pm. Everyone went shopping for groceries. Eileen and I walked to the nearby supermarket (the Parisian) for more food. I bought candy bars and rice pudding. Before we went back to Tusitala for the trip back to Las Hadas the cruisers on three boats decided to buy ice for their iceboxes. At the back of the restaurant they sold the largest blocks of ice we had ever seen: 55 kilos (120 lbs) enough to fill a bathtub. The ice was probably intended for the holds of fishing boats to keep fish fresh. The men had to break the block into three chunks to transport it. With one man on either side of a large canvas bag they carried each piece down the ramp to the dinghies. Although we didn't use ice, Dan helped carry the ice for the most elderly cruiser, Bob, who was in his 70s. It took a couple of trips by dinghy to load all our purchases into *Tusitala*.

*Tusitala* was packed to the brim, with most of the passengers in the cockpit, all our groceries in the main cabin, and one block of ice in Bob and Marge's icebox. We left the remaining two huge blocks of ice in one of the inflatable dinghies we were towing, and covered it with a sheet to shield it from the hot sun.

There was quite a bit of chop as we motored *Tusitala* back to Las Hadas. During the one-hour trip we were alarmed to see saltwater lapping over the sides and into the bottom of the inflatable dinghy carrying the two blocks of ice. As we glanced back the lumps under the sheet appeared to be diminishing in size, although we couldn't tell how much. After re-anchoring at Las Hadas, Bob pulled the inflatable dinghy alongside and removed the sheet. There were two pathetic chunks of ice about the size of bricks sitting in a large pool of water. Bob generously shared his 40 lb piece of ice in his icebox by breaking it into thirds to give to Eddie and Eileen on *Nubian*, and Bill and Michelle on *Akavit*. We couldn't stop laughing at our stupidity. How could we think ice wouldn't melt in the warm saltwater? It was a lesson in how not to carry ice in a hot country. We clearly had much to learn about cruising in Mexico.

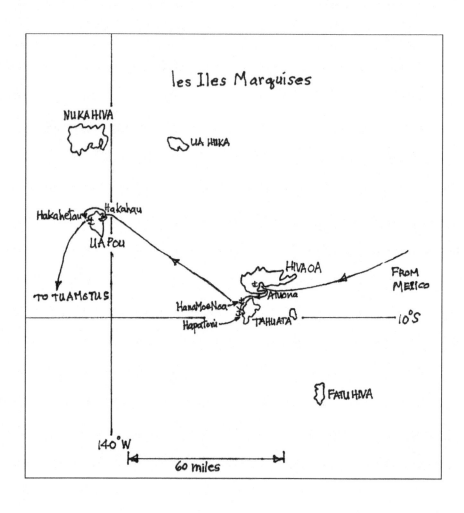

Cruising the Marquesas in Shaula

May to October 1989

# 4 FRENCH POLYNESIA

### Landfall in Paradise

After five months cruising in Mexico we reached our last Mexican port, Puerto Vallarta. Puerto Vallarta is the usual jumping-off place for the Marquesas, and there we began the frantic activity of getting the boat and ourselves ready for the passage. We joined with many cruisers who were also preparing to sail to the Marquesas, making frequent trips to the markets for provisions, and to restaurants in Puerto Vallarta for inexpensive Mexican meals. While there Dan discovered oil in the engine's freshwater cooling system. It took a week for him to diagnose the problem (a leaky head gasket), send for a new head gasket from the States, install it, and put the engine back together. We spent the mornings varnishing, putting six or seven coats on all the exterior wood. After filling up with fuel and water we were finally ready to go.

We departed from Puerto Vallarta on April 3rd with the intention of stopping at Socorro Island, a Mexican navy outpost 300 miles off the coast of Mexico. The sailing was much more pleasant than sailing off the Washington coast, with sunny days, light breezes and slight seas. On the third day out we could see Socorro Island very clearly at sunrise. We entered the harbor to anchor in early afternoon. Our friends Michael and Sheila on *Kantala* greeted us on arrival. The commandant of the Navy base, Fernando, was extremely friendly to visiting cruisers. He swam out to the boats every afternoon to visit the cruisers and invited us to his office for coffee and conversation the next morning. Michael had succumbed to a severe case of food poisoning after their arrival, but was on the road to recovery. He received excellent care from the Navy doctor, who came to the boat daily and gave him IV antibiotics. We snorkeled in the warm water in the harbor, seeing hundreds of colorful fish. We took a tour of the impressive farm where they raised chickens, pigs and rabbits, and the bakery, where they baked fresh bread daily. We were able to buy fresh eggs and bread for our passage to the Marquesas.

On our second day there I came down with a sore throat, which developed into laryngitis and a hacking cough after several days. I had to decline the offer of a jeep trip on the island, but Dan enjoyed the trip up the steep slopes, and a hike to the volcano on the top of the island with two Mexican navy guides. They had magnificent views of the island from the top. After six days, when we visited the commandant for coffee, I still had a hacking cough. The commandant insisted that the Navy doctor examine me. The doctor checked my throat, and gave me an injection of ampicillin, and tablets to be taken for a week, all for free. I was very appreciative, having learned by experience that standing watches on a passage is not a good way to get over a cold or a cough.

We departed the next day for the 2,400-mile passage to Atuona. This was idyllic sailing—light to moderate northeast trade winds and slight seas. We watched spinner dolphins leaping through the waves, schools of flying fish soaring across our bow, a large sea turtle, and lots of different seabirds, including an occasional booby. We soon adjusted to the motion and could sleep on our off-watches. Every evening we checked into the Pacific Maritime net to report our position and learn where our friends on other boats were. Occasionally after roll call the net controllers, based on land, would set up a phone patch to friends and relatives. We really enjoyed talking to Dan's family in California and my sister and brother-in-law and friends in Seattle. We were thrilled to learn that Dan's LWOP (leave without pay) had been extended for another year.

A week after departing we crossed the inter-tropical convergence zone, renowned for squalls, heavy rain, and periods of doldrums. We repeatedly passed under small, black clouds and were hit by strong gusts of wind, and sometimes torrential rains, then came out to clear skies and moderate winds again. It was frightening at first, but we found the winds were less than 25 knots, and the squalls lasted only 15 or 20 minutes. Fortunately we got through this in only 24 hours, and then we were in the southeast trades, with steady winds, rolling seas, and ideal downwind sailing.

Two days later we crossed the equator for the first time, an exciting event for us. Although we're not superstitious, when we crossed the equator we followed the well-known tradition of toasting King Neptune to assure a safe journey. For our toast we poured a jigger of rum and crumbled some cookies into the ocean. Just after crossing the equator we caught a 30-inch tuna, which made three delicious meals. When that was consumed we caught another large tuna, 32 inches. The tuna meals were very welcome just when we had run out of all fresh food. Twenty days after departing Socorro Island, we spotted land just at sunrise—beautiful, jagged peaks, waterfalls, and everything unbelievably green and lush. In early afternoon we dropped our sails and motored into the harbor to anchor among the 14 other boats. We were exhilarated, but also exhausted. Our friends Dick and Cris rowed over to our boat with some cold white wine and beer, a most welcome treat. We slept like contented babies that night in the calm anchorage.

After recovering from the passage we were invited to a landfall party the next afternoon to celebrate our arrival with friends from five other boats. We consumed lots of great food—smoked oysters, clams, salami, olives, French bread and wine. Our happy hour was followed by a delightful bluegrass session with Dick on banjo, Cris on mandolin, Greg on guitar, and me on fiddle, playing and singing the many tunes we had practiced in Mexico.

Dan and I had made the long passage without any major problems. This was completely different from our cold, wet Seattle to Hawaii trip, where we had gear failure, gale force winds and fog. Landfall in the Marquesas amongst many sailing friends was delightful. Now I understood why so many people crossed oceans in small boats. This passage transformed me into an enthusiastic ocean sailor.

## Bluegrass in the Marquesas

After our wonderful 20-day passage from Mexico to the village of Atuona on Hiva Oa we continued to get together for daily bluegrass sessions with Cris and Dick on *Chatauqua*. A teacher at Dan's school, Sandy, told us before we left Seattle to be sure to look up Julia when we reached Atuona. Sandy had met Julia during a summer trip to China. Julia was a teacher from France who taught art at the elementary school in Atuona. A few days after our arrival Cris and Dick joined Dan and me to search for her. We trudged up a narrow path to the top of a steep hill near the anchorage. Julia lived in a beautiful European-style stone house with her two young children. We had a delightful visit with her and she invited us to come to dinner and to play bluegrass that evening.

In late afternoon the four of us again hiked up the hill from the anchorage with our instruments and joined Julia and several of her teacher friends and their children for a jam session of bluegrass music. Afterwards Julia served a delicious dinner of roast lamb from New Zealand and local fruits and vegetables. It was a fabulous evening of chatting, singing and playing music, and learning how the French ex-pats lived on this remote island.

After our stay in Atuona we sailed to a small nearby island and anchored at Hana Moe Noa. It is a beautiful bay with a white sand beach and turquoise water surrounded by acres of coconut palms. No one lived ashore, but we met Aka and his father who were there to harvest coconuts for copra. The day after we arrived we watched Aka and his father construct a small hut out of woven coconut fronds near the beach where they would stay until the harvest was over. They built a long wooden table and benches alongside the hut for their meals, and in the evenings lit a campfire for cooking. Every morning Aka would go out to gather coconuts. Dave, from the sailboat *Silverwind*, and Steve from *Tandem Cay*, the two youngest men in our group, offered to help. They rose early in the morning and spent days with Aka in the hot sun gathering coconuts, husking them with machetes, and bringing them back to camp to dry them in the sun on raised platforms. It was backbreaking work. After weeks of drying, the dried coconut meat would be removed from the shells, packed in burlap bags, and shipped overseas to be used in manufacturing soap and cosmetics.

In the evenings Aka and his father invited us over for dinner at their campsite. They were amazingly good cooks. They prepared feasts of poisson cru (bite-size pieces of raw fish marinated in lime juice and coconut milk), roasted breadfruit with a creamy garlic-coconut sauce, and tropical fruit, all served on platters of large banana leaves. One afternoon Aka shot a goat while out gathering coconuts, and invited us all for a dinner of goat roasted on a spit, which was a real treat. There was no clean up—we ate everything with our fingers and threw the banana leaves into the bushes when we finished. Dinner was always followed by a session of bluegrass music. Aka was an accomplished guitar player. With a little coaxing and a few swigs of rum, he entertained us with some Tahitian and Tuamotuan songs.

We sailed on to the next anchorage, Hakahau, on the island Ua Pou. It was the end of May, and Dan and I were looking for a place to celebrate our 21st wedding anniversary. We found a small bamboo hut in the village that served hamburgers and Heineken. Dick and Cris, Steve and Heather, and Dave and Debbie joined us. During our lunch we met the director of education, Toti, who was having lunch with his friends. When Dick mentioned we had a bluegrass band, Toti invited us to give concerts the next day at the kindergarten and the high school. The next morning we gathered in a small pagoda in front of the elementary school to play for the pre-school children. The young kids really got into the spirit, at first clapping along to the music, then dancing around in a circle inside the pagoda, in perfect rhythm. At noon we gave a short concert of bluegrass tunes next to the soccer field while the high school students were on lunch break. This was not as successful. A few students listened to us, but soon drifted off to eat lunch. We had the impression they were more interested in playing soccer, or perhaps listening to rock music.

When the concert was over Toti treated us to lunch that was prepared for us at the restaurant where we had met. It was a delicious feast of cooked crab, roasted breadfruit, fried bananas, French wine and coffee. After lunch Toti took us to the elementary school where he used to teach. We played bluegrass for a half hour to an enthusiastic group of five- and six-year-old children. They particularly liked "Pig in a Pen" and "This Land is Your Land". Cris and Dick had cleverly replaced the line "from California to the New York island" with the names of four Marquesan islands, so it became "This land is your land, this land is my land, from Ua Pou to Nuku Hiva, from Fatu Hiva to Hiva Oa, this land is made for you and me." Big smiles broke out on the children's faces. They loved it.

The next day we were invited by the director to play at an all-day children's dance festival. The kids were adorable, all dressed in their costumes of grass skirts, with flower leis. There was a competition for the best dancers for the girls and best muscle-man for the boys. We were amazed at how the tiny girls could swivel their hips to the sound of the drums, in perfect imitation of the adults. The young boys flexed their arm muscles, which were almost non-existent, in He-Man poses. During lulls in the program we played a few bluegrass tunes to fill in.

Our last anchorage before leaving the Marquesas was Hakahetau, just a few miles from Hakahau on the island of Ua Pou. There we met the mayor, who taught a class of five- and six-year-olds. He had heard about our bluegrass playing from friends in Hakahau, and wanted us to play for his class. We played in the classroom to a very attentive group of children. The mayor introduced each piece before we played, translating the title into French. We couldn't conceal our laughter when he introduced "Whiskey before Breakfast" as "Aperitif avant le petite dejeuner"—it sounded so elegant in French. Afterwards the mayor introduced us to a friend of his, Alphonse, who invited us to his house for lunch. We watched him cut up a beautiful, large turquoise parrotfish to prepare poisson cru. This fish had become one of our favorites to observe when snorkeling on coral reefs in the Pacific. We hated to see it being cut into small pieces, but we enjoyed eating it anyway. Alphonse was a great guitarist, and was fascinated by Dick's banjo. After we played some bluegrass Alphonse's father borrowed Dick's banjo and laid it out on a sheet of plywood to make a pattern. He was determined to make a banjo for Alphonse. We thought it would take some ingenuity, but he would probably succeed.

In the space of a few weeks Dick, Cris and I had played bluegrass music at five schools and at many informal jam sessions with the local Marquesans. In exchange the locals sang their songs, often accompanied by guitar. I was certainly glad I had brought my violin along on our boat and grateful that we had met Cris and Dick, who were accomplished musicians and were able to sing in beautiful harmony. In these islands where we had difficulty communicating with our limited knowledge of French or Marquesan, music was a great way to communicate and a source of much delight and entertainment.

## Black Pearls

Our five-day passage from Ua Pou in the Marquesas to Takaroa in the Tuamotus was slow and easy, just the way I like it. We had enough wind to sail most of the way, and the seas were slight. The

Tuamotus are a group of 70 atolls spread across a vast area of ocean 1,000 miles by 2,000 miles, on the path between the Marquesas and Tahiti in French Polynesia. Each atoll consists of a ring of coral topped with many small islands, called motus, surrounding a deep lagoon. Because the atolls are low-lying they are only visible at a distance of about 10 miles. Strong currents around the atolls have caused many ships to wreck on the shores, hence they are known as the dangerous archipelago. Takaroa is at the edge of the group of atolls, so we would not have to pass through a maze to get there. Our satellite navigation stopped working while we were still in Mexico, so we had sailed from Puerto Vallarta to Atuona using celestial navigation. Cruisers we met in Atuona whose Sat Nav had failed suggested a trick to get it working again. They turned it on, re-initialized it, and waited for days for it to get a fix. (Normally, Sat Nav would obtain a fix every three or four hours). Before we left for the Tuamotus we tried this trick, and after 24 hours heard the beeping indicating we had a fix. After that it continued to give fixes regularly. Having a working Sat Nav made the passage much safer than when using only celestial navigation, especially when the sun is obscured by clouds, or at night. It provided us with the frequent fixes needed where strong currents sweep the boat off course.

At dawn on the fifth day of our passage we spotted the tops of coconut palms on Takaroa when we were nine miles off. The pass into the lagoon at Takaroa is a good one for anyone with the patience to wait for a slack current. The maximum current is nine knots. Since our maximum speed under power is five knots we had to wait until the current was less than five against us in order to power through it. We followed the rules for predicting slack tide, but this is an inexact science. When we approached the pass and looked in towards the lagoon we could see there were no large waves, indicating it was close to slack tide. When we entered the pass the current was three knots against us, reducing our speed to about two. The pass is short so we were soon in the calm waters of the lagoon, where we could anchor and rest up from the passage. We didn't know where in the lagoon to anchor, but Dan picked a nice-looking spot across from the pass that had a sandy bottom, shelter from the prevailing winds, and no dwellings in view. This turned out to be a very lucky choice. Steve and Heather on the 27-foot *Tandem Cay*, had planned to sail to the atoll of Manihi, but couldn't reach the pass before sunset, so changed their plans and sailed to Takaroa just behind us. Dave and Debbie on *Silverwind* planned to sail to Takaroa, and arrived about an hour later when it was actually slack tide. They had no problem motoring through the pass. *Tandem Cay* and *Silverwind* anchored near us in the lagoon. Of all the sailboats that had left Mexico that season to cross the Pacific we were in three of the smaller boats, ranging in length from 27 to 31 feet. (Most of the sailboats were in the 35-to-45-foot range). We launched our dinghy and rowed over to invite our four friends to a landfall party on our boat later that afternoon.

After a passage I'm anxious to get off the boat as soon as possible, so I rowed along the motu for a half mile and spotted a house on shore at the next motu. Several people were out in front of the house and a young man motioned for me to come into their dock. He introduced himself, Alfred, his wife, Pauline, their two young children, his father, and a friend about his age. Pauline was Chinese, but had spent six months in New Zealand learning English, so we were able to communicate. They gave me a very warm welcome, and showed me around their property. They had two dozen chickens, and a manmade lagoon to temporarily hold fresh fish they had caught. They informed me that the rows of buoys in the lagoon in front of their house were part of a black pearl farm. They showered me with gifts, including a fresh drinking coconut, two beautiful cowry shells, two large black oyster shells, a dozen fresh eggs, and a large fresh fish. They insisted on taking me back to our boat in their 24-foot runabout, after putting our small dinghy on the foredeck, so that I wouldn't have to row back. Dan and I divided up the eggs to share with our friends, and we invited them to stay for a fish dinner after happy hour on our boat. We had a fantastic landfall party and dinner, and were all delighted that we had stumbled upon a family who were culturing black pearls.

The head of the family was Ah Sahm, about 48 years old, who was a quarter Chinese and three-quarters Tuamotuan. He had retired from a job as supercargo on one of the copra boats. Ah Sahm worked on culturing pearls with his son Alfred, who had been a representative for a paper company in Papeete, and his son's friend, Hanere from Moorea. Alfred's wife, Pauline, took care of their children, Nancy and Tony, and helped with cooking and household chores. With Pauline's English and Dave's knowledge of French from living in Canada, we managed to communicate fairly well. They told us that they had watched sailboats go by over the years they'd lived there, but none had ever stopped and anchored near their property. They were delighted when we showed up. The family gave the six of us cruisers honorary Tuamotuan names, but Dan and I were always called Momi and Popi because we were the senior citizens of the group (at age 55 and 56) and these were terms of respect. Within a day or so Steve and Dave were spending almost all day working on the pearl farm with Ah Sahm, Alfred and Hanere. The rest of us spent many hours at their house, helping with the work and learning about life on a well-run cultured black pearl farm. During breaks we went fishing in the lagoon from the family's runabout, catching many fish for our dinner. Ah Sahm frequently invited us to dinner with the family, sometimes potluck, and we enjoyed some wonderful food. Fish and rice were a big part of most meals. The fish was fixed many different ways—fried, barbecued, broiled or marinated to make poisson cru—all delicious. We had some hilarious times swapping jokes and stories.

We found the black pearl farm fascinating. The native oyster has a dark grey iridescence on the inside of its shell, and the pearls that are formed inside become beautiful shiny black pearls. The oysters are drilled with a small hole near their hinge, and hung by six foot long vertical cords onto a horizontal rope stretched between two buoys anchored in the lagoon. The oysters are kept at a depth of 10 to 30 feet. Every few months each oyster must be cleaned, as mussels, barnacles, and other marine growth soon cover the shells and weighs them down. Each morning the men dive down and check the oysters, bringing in 200 or more oysters at a time to be worked on. They bring these onto the dock in front of the house in buckets and clean them by scraping off the marine growth gently with the sharp edge of a cleaver. They remove the remainder using a scrub brush. To keep the oysters alive they never keep them out of the water for more than three or four minutes. Ah Sahm had hired an expert from Japan to seed the oysters in December. The seeds are smooth, round spheres made in Japan from imported Mississippi mussel shells. If the oysters had been seeded the previous December, Alfred checks to see if they still had the seed. He had a special tool to pry open the shells about a half inch so he could see the gonads where the seed was placed. Only one out of five of the seeded oysters keeps the seed. If the seed was gone the oyster was placed with those oysters to be seeded the following December. Ordinary cultured black pearls are worth about $50 each, while some of the larger ones with blue, green or reddish highlights are worth $200 or more. With lots of hard work, a farm could grow many thousands of oysters, and produce an annual income in the hundreds of thousands of dollars. The family had just started the farm three years before, and it takes up to five years to produce a good size pearl, so they hadn't had a harvest yet. It was exciting to see their operation and their enthusiasm, and we were all hoping for their success.

When we weren't working on the oysters or eating sumptuous meals with the family at their house, we swam in the warm water where our boats were anchored, or snorkeled to look at the colorful fish. Using hermit crab tails for bait we caught several large, exotic fish from our boats, and shared them with our friends and the family.

After eight days Dan and I decided to leave for Ahe, another atoll in the Tuamotus. The evening before we left Ah Sahm invited everyone to a going-away feast for us. I made a three-bean salad, Heather made a spaghetti dish, and Debbie baked a cake. The family served us poisson cru, chicken, potato salad and French bread. After dinner we played Quilcene, a dice game that we had learned in Mexico from friends on a Seattle boat. It's a game in which success depends on the willingness of the players to take risks. Ah Sahm caught on very quickly and won most of the games. Our cheers and shouts of delight whenever anyone made a big score probably woke up the two children and echoed across the lagoon to wake the neighbors.

Before we left the next morning Ah Sahm invited Dan and me to his house for coffee and warm French bread, and gave us parting gifts of two huge oyster shells with embedded black pearls, some ramboutons (a delicious fruit that tastes like lychees), and a dozen fresh eggs from his chickens. It was very difficult to leave. Our visit to Takaroa had been the highlight of our trip. There were no hotels, bed and breakfasts, or restaurants on the atoll. The only way we could have had this experience was to sail to Takaroa, bringing our own accommodation, food, and water.

## The Music of Ahe

After our visit to the pearl farm at Takaroa we were off to Ahe, another small atoll in the Tuamotus. It was only an overnight sail to this atoll, and we arrived at dawn on June 30$^{th}$. By the time we waited until slack water so we could motor through the pass it was mid-afternoon. We crossed the lagoon to anchor in front of the village where three boats were anchored, including *Chautauqua*, with our friends, Dick and Cris, on board. Another sailboat, *Kantala*, with friends Michael and Sheila aboard, was tied up to the wharf.

This small atoll was home to many talented musicians. Most evenings a group of villagers gathered at the wharf after dinner to sing and play music. Hiti, a good-natured 400-pound local, sang and played a homemade ukulele. A cruising sailor had given him a wooden cigar box, which he fashioned into a ukulele by adding a neck, frets, carved wooden pegs, and nylon fishing line for strings. Another villager played guitar, and everyone in the village sang. Cruisers from the anchored boats rowed into the dock every evening to join in the music. Cris, Dick, and I brought our instruments—a mandolin, banjo, and violin—to sing and play bluegrass tunes. One night we played till 10:30 when we noticed Hiti and his family had fallen asleep on the wharf. They and two other families slept there all night. On Saturday evenings the locals brought their instruments and electronic equipment to play dance music in the Town Hall. The locals danced a modified two-step, while the visitors spiced it up with some modern variations. Sunday afternoons the locals set up their small gasoline generator, amplifiers, and speakers to give a concert in the gazebo near the wharf. They entertained us with Tuamotuan and Tahitian songs. Sunday evenings everyone gathered for a service at the LDS (Latter Day Saints) church. We enjoyed listening to their beautiful and enthusiastic hymn singing.

During the days we swam and snorkeled in the clear, calm waters of the lagoon. Hiti was an expert fisherman, and often went out to the pass to spear fish. One afternoon he and a fellow from one of the Swedish boats came back with a variety of fish, enough to give one to each of the seven anchored boats. He gave us a grouper, which made excellent poisson cru. Other times we had potluck dinners on our boat, *Kantala*, or *Chautauqua*.

The children of the village were on school vacation, and spent most days visiting the boats at anchor by swimming or paddling out on the surfboard that Michael and Sheila loaned them. Groups of kids would climb into our cockpit and entertain us by singing and teaching us a few words of French. If they arrived at lunchtime I always made them big batches of popcorn, which they devoured enthusiastically. If they stayed too long and we wanted some quiet, we said "Off the boat", and they dove off and swam or paddled back to the wharf. Michael and Sheila had taught them these few words of English. Their boat was tied up to the wharf so that Michael could restore a beat-up dinghy they had acquired in Mexico. The children could easily climb aboard their boat, and they had as many as 20 children at a time visiting.

On the fifth day of our stay at Ahe Michael finished re-building the dinghy he had worked on so diligently for the last several weeks. He and Sheila invited all the boaters and the villagers to a launching party on the wharf. We spent the morning preparing Kool Aid, large bowls of popcorn, and homemade cookies for the launching. The children ate first, contrary to Polynesian custom, and they were ecstatic. They filled their bowls with popcorn and cookies to be sure they would get their share. Hiti's mother-in-law, Ela, said a solemn prayer to bless the dinghy. Cris, Dick, and I played bluegrass tunes for the celebration. The children were fascinated by my violin, and wanted to try playing it. When I told them the bow was strung with hairs from a horse's tail, they all wanted to feel it. We finished our playing with an Irish tune—the "Irish Washer Woman". Michael performed an excellent Irish jig on the wharf to everyone's delight. At last Michael and Sheila lifted the dinghy and tossed it off the wharf down into the water. Unfortunately, the mast broke in two on impact. We all collapsed in laughter. Michael was unfazed—he could repair the mast, and even without it they had a fine rowing dinghy.

After the launching Dan and I, Michael and Sheila, and Cris and Dick were invited to dinner at Emilie and Tetua's house. They had prepared a fantastic feast of sashimi with raw tuna, poisson cru, barbecued tuna, chicken, rice, and spice cake. Following tradition the children left while we ate, and returned to eat the leftovers after we finished. After dinner Emilie and Tetua played us a tape of Tahitian songs sung by Emilie's three nieces. It was one of three tapes they had made, which they sold in Tahiti for $16 each. We were impressed by the extraordinary musical talent of the people on Ahe. After dinner Cris, Dick, and I started playing bluegrass music, but after a few minutes my bow quit working completely. When I bowed no sound came out. Looking at it closely, I realized the hairs were coated their entire length with butter. I should never have allowed the children to touch the hairs on my bow, especially after they had eaten buttery popcorn.

The next morning Michael proudly rowed his new dinghy around the anchorage. He came a little too close to our boat. Unable to see behind him, he ran his dinghy's mast smack into our bowsprit, which sticks out seven feet from our bow. Down came his mast for the second time. He would have to repair it again. Meanwhile, I had the task of restoring my bow if I wanted to continue playing my violin. I unscrewed the end of the bow all the way so I could remove one end and wash the hairs without getting the wood wet. I gently swished the loosened hair in a detergent solution in our dishpan, rinsed it thoroughly, and hung it out to dry in the cabin. I had never tried this before, but it worked. The next day I rubbed rosin on the bow hairs, and I could play the violin again. Michael was able to resurrect the mast on his dinghy, so now had an excellent sailing dinghy.

A couple of days later we planned to leave for the three-day sail to Tahiti. Bastille Day, July 14, was coming up in less than a week, and we wanted to be there for the big celebration. In the morning before our departure a large copra boat came into the lagoon with supplies. We boarded one of the runabouts they use to ferry people out to the big boat. With our list in hand we waited our turn to step into the hold to purchase supplies—two boxes of crackers, five baguettes, a can of dried milk from New Zealand, and three kilograms of New Zealand apples. When we departed that afternoon, along with *Kantala* and three other boats, we were well supplied with fresh food. We left with fond memories of beautiful music wafting in the air, wonderful meals we had shared with friends, and the charming children of Ahe, who loved singing and eating popcorn.

## Penniless in Polynesia

Our problem started in Atuona, our port of arrival in the Marquesas. Before we left Mexico we had written to our mail service in Seattle to send our next mail package to the Atuona post office. When we arrived in Atuona we immediately went to the post office to pick up our mail package, but it wasn't there. This was disappointing. It was critical that we receive our mail, because our MasterCard had expired at the end of April. Our new cards should have been in our mail package in Atuona when we arrived the first week in May. Without our new credit cards we would have no way to pay our bond. The bond is equivalent to airfare back to a person's home country, in our case $850 each for a total of $1,700. The bond is to assure that boats leave before the beginning of the hurricane season in October. It is intended to save the French government from costly rescue of boats that are caught in hurricanes. The government returns the money in full when you check out of the country before departing. After asking about our mail every few days at the post office, we gave up. Assuming it was a case of miscommunication, we wrote to our mail service again before we left Atuona, asking them to send our mail package to the post office in Papeete.

After arriving in Atuona we had 15 days to pay our bond. Most cruisers use their credit cards at the bank. With an expired credit card we were forced to pay our bond with the $1,100 in travelers' checks plus $600 cash we had on board. This left us essentially penniless. Fortunately, our friends Cris and Dick offered to loan us $200 so we could buy food as we travelled through the Marquesas and the Tuamotus. We hated having to borrow money, but we knew we could pay it back when we reached Papeete and picked up our mail package. Along the way we spent the $200 Cris and Dick loaned us and asked for another $100, which they gladly loaned us. In the Marquesas and Tuamotus, we were able to live on a lot less than the $1,000 per month we had budgeted. There was little to buy, and the locals often gave us local fruit such as pamplemousse, lemons, mangos and bananas.

After two months of cruising through the islands, we reached Papeete. Because the harbor was full, we anchored at Maeva Beach, where anchoring was free, and where many of our friends were. The morning after we arrived we took the 20-minute ride in Le Truck into Papeete to pick up our mail. The bus ride was an adventure in itself. Le Trucks are small, brightly decorated buses with hard bench seats that are used mostly by the locals. During the ride loud Tahitian and American music blared from huge speakers in the front of the bus. We walked to the large post office. No mail! It was extremely frustrating. Papeete was a major tourist destination with lots of temptations, but we had almost no money to pay for any of it.

In downtown Papeete the streets are lined with fine, but expensive French restaurants which we avoided. Every night near the harbor there were dance competitions, in which dance teams from different villages performed their traditional Tahitian dances in native costumes. Tickets for the nightly performances cost $25 each, so we wouldn't be able to attend these.

Within walking distance of our anchorage at Maeva Beach sat a huge department store, the Euromarche. Every few days we strolled through the store gazing in amazement at the food shipped in from all over the world—broccoli, celery, tomatoes, chicken, and beef, all at prices about five times what they cost at home. By selecting only local foods we were able to buy a few items such as potatoes, carrots, papayas and pamplemousse at reasonable prices. Instead of dining at the fancy restaurants in Papeete we returned to the boat, where I cooked dinners consisting of canned pork and beans, sardine spaghetti (from sardines bought in Mexico), bean casseroles, or chowder made from dried tuna we had caught on the way to Papeete.

It was extremely disappointing that our mail hadn't arrived, and therefore we had no credit cards to get cash from the bank. On our next trip into town we walked to the parcels post office several blocks away. Since it had been three months since we had received mail, we thought our mail service might have put it in a box, rather than the usual large envelope. When we didn't have mail at the parcels office, we decided to call Barbara and Tom in Seattle to see if they could wire us $3,000. Before leaving Seattle we had given Barbara power of attorney and added her name to our bank account. Barbara readily agreed send us the money. She said she would also check with our mail service to find out if or when they had sent our mail.

The third day after our arrival was Bastille Day, celebrated by a large French military parade, followed by a parade of the local people. We joined Cris and Dick for a buffet breakfast at Maeve Beach Hotel. After having seconds and thirds, we took the bus into Papeete. By the time we arrived the French military parade was over, but we enjoyed watching the groups of locals from different villages in their colorful costumes, singing and dancing as they paraded down the street. In the afternoon we wandered into the free arts and crafts building to see the native shell jewelry, carvings, and tapa cloth.

When we were in town we watched the canoe races in the harbor in the mornings. We ate lunch at the small vans near the harbor, where a Chinese lunch cost only $5. We visited the free Black Pearl Museum with its description of how black pearls are cultured in French Polynesia, and beautiful displays of jewelry made of black pearls. The Maeva Beach hotel had Tahitian dancing every evening in the restaurant, which was free with the purchase of a glass of wine. One evening they were so busy that the drinks waitress never got to our table to take our order, so we saw the dance program without the obligatory wine.

Papeete was full of cruising boats, which gather in the harbor for the week-long Bastille Day celebrations. When not being entertained by the delights of the city we joined in potlucks or happy hour with our friends on boats at anchor or in the harbor. One late afternoon

Michael and Sheila invited us all over for wine on their boat. Our happy hour lasted four hours, and then it was time to return to our boat and make dinner. Another day Cathy and Joe, on their boat *Champagne*, had us and Cris and Dick over to their boat for a potluck. They had been moored at the wharf since June, ate out at French restaurants, and attended the dance competitions every night. It was costing lots of money, but they were thoroughly enjoying Papeete. We met Pat Henry, who was single-handing on *Southern Cross*. She arrived in Papeete with only $3 left in her cruising kitty, and was spending her time painting miniature watercolors, which she sold to cruisers for $30 each. By comparison to her we felt rich.

When we went back into Papeete three days after talking to Barbara in Seattle we called her again. She was a fantastic business manager. She had gone to Pike Place Market and picked up our mail for us, sorted through it, and found a cancelled check to our mail service for $100. Our mail service had cashed our check but not credited it to our account. They had stopped forwarding our mail because they thought our postage fees had run out. Barbara said she would wire money from the main branch of the Seafirst Bank in downtown Seattle to the Bank of Tahiti the next day. She would send our package of mail to the Papeete post office. We were overjoyed—at last we would have money and mail. However, we were terribly disappointed that our mail service had failed in the one job we had signed up for, forwarding our mail to us.

It took five days for the money to arrive from Seattle to the bank in Papeete. Life didn't change much now that we had money. We had learned to enjoy Papeete without spending much, but we were glad to have enough to buy food for the next three months as we visited the other Society Islands of Moorea, Huahine, Raiatea, and Bora Bora.

The next morning we treated Dick and Cris to a sumptuous buffet breakfast at the Maeva Beach Hotel and returned the $300 we owed them. We really appreciated their generosity. On our next trip to the

Euromarche we bought frozen chicken imported from Mississippi, a real treat after our mostly canned food. A week later, just before departing Tahiti for Moorea, we picked up the large package of mail at the Papeete post office that Barbara had sent us. Our new credit cards were inside. Of course, for now we didn't need them.

What did we learn from being penniless in Polynesia? It's wise to keep ample cash on board the boat. If things go awry it's important to have good friends to help out. We also learned it's possible to be entertained in Papeete with very little money.

## A Visit from Dan's Brother

In August 1989 after our stay in Papeete, Tahiti, Dan's brother, Gary, accepted our invitation to visit us in French Polynesia. He booked a six-day package deal at the Club Med Resort on Moorea, and then would join us on the boat for the second week. We were excited to have our first visitor on the boat since leaving Mexico.

The day before Gary was due to arrive we motored over to the bay in front of Club Med to anchor. The bay was bustling with noisy outrigger canoes powered by outboard motors, glass-bottom boats, and powerboats pulling water skiers. A two-knot current roared past our hull. It was the worst anchorage we had experienced in French Polynesia. The next morning we rowed the dinghy ashore and went to the front desk at Club Med hoping to visit with Gary over a cup of coffee at the café. However, the clerk at the front desk told us that the only way we could enter the premises was to book a lunch or dinner at the restaurant. Lunch cost $50 per person, which was way beyond our budget, so we declined. We returned to our boat and waited for Gary to contact us.

Gary arrived that afternoon in a large Club Med outrigger canoe filled with snorkelers. The captain pulled the outrigger alongside our boat and Gary climbed into our cockpit. The airlines had lost one of his bags with all of his clothes, so he was wearing clothes he had intended to give Dan. We had a nice visit until late afternoon when the outrigger came by to pick Gary up and return him to Club Med.

The next morning Gary came aboard our boat again from the outrigger, this time bringing a bag of goodies we had asked for—four new cockpit cushions, two pairs of swim shorts for Dan, a battery for our voltmeter, a repair kit for our marine toilet, and a flash attachment and film for our camera. It was like Christmas! These were items that were not available or prohibitively expensive in Tahiti. We packed a picnic lunch and had an enjoyable afternoon snorkeling with Gary on a nearby motu. The Club Med outrigger canoe picked Gary up at the motu to return him to his hotel in late afternoon. Gary told us that since he had all his meals and plenty of entertainment at Club Med we should move to a better anchorage, and he would join us when his stay there ended in five days. So the next morning we motored a short distance to Opunohu Bay, and enjoyed a much calmer and quieter anchorage.

While at Opunohu Bay we had time to write letters, wash our clothes in buckets on the wharf, fill up our water tanks and buy groceries. Four days later when Gary arrived at the park on shore, along with his luggage, Dan picked him up in our dinghy. We had lunch and dinner aboard our boat. Since our quarter berth was full of gear that we needed for our two-year offshore trip, Gary had to sleep in the cockpit. He laid out the four new cockpit cushions on the narrow cockpit seat for a mattress. He used Dan's sleeping bag, which we had stored, in the quarter berth, but he turned it inside out, because we found it was moldy inside.

The next morning we walked down the road to the Moorea Lagoon Hotel to watch the end of the second annual Moorea marathon. Steve and Heather joined us. Runners from all over the world ran 26 miles along the road of coral rubble that circles the island. We watched the winner, a small Japanese man, finish the race and collapse just after crossing the finish line. He seemed to be in considerable pain. He was carried on a stretcher into the hotel to recover. Two Tahitian men took off their shoes and finished the race barefoot. We were surprised to see several cruisers we had met earlier running in the marathon. They had had only three weeks of training. A tall blond man from a 60-foot luxury sailboat won the half marathon. We were impressed that these cruisers were in such good shape.

We had hamburgers for lunch at the small café, and attended the award ceremonies at the Bali Hai Club in the afternoon. A troupe of Tahitian dancers put on an entertaining show for the marathoners. We had skipped the weeklong dance festival in Papeete because of its high price of $25 per person each night. Now we were able to see some excellent dancing for free.

The next day was a work day. Dan and Gary took their clothes and Dan's sleeping bag ashore in buckets to wash them on the wharf, where there was a supply of water from a hose. We had wine and dinner on board *Shaula* that evening. The next day we walked to the Moorea distillery, where we had a free tour, watched the extraction of pineapple juice, and tasted a variety of delicious liqueurs, including coconut, ginger, pineapple and papaya.

The weather, which had been blustery, calmed down enough that we could sail to the next island of Huahine, only 85 miles away. We left in mid-morning for an overnight sail, hoping to arrive after sunrise the next day. It was a pleasant sail, with gentle winds and flat seas. Dan and I stood three-hour watches, Gary joining us in the cockpit for part of our watches. At night Gary slept in the bunk of whoever was on watch. We made good time, and could see Huahine at first light. Gary woke up feeling queasy, and had to take a Bonine tablet for his seasickness. As we approached the entrance to the bay I woke Dan so we could drop the sails, and motor through the pass into

Fare. We anchored amongst many yachts in the deep water bay—a beautiful harbor with lots of inter-island ships at the large wharf. The best entertainment on Huahine was watching all the people and goods arriving and departing at the wharf. All of the islands are supplied from ships that come from Papeete.

Our toilet plugged up the next day with crust from saltwater building up in the tubing. We had to use the shore facilities. Dan and Gary took the toilet ashore and spent the afternoon cleaning out the tubing and replacing the worn-out mechanical parts. Dan was able fix it and re-install it in the boat. It was unfortunate this problem coincided with Gary's visit. He was certainly learning about the difficulties of living on a small boat.

The next day Gary moved off our boat after finding a room at the nearby Huahine Hotel. It was a nice, small hotel, the only difficulties being that the owners spoke only French, and mosquitoes swarmed around the dining room and bedroom at night. When Gary complained to the staff, they gave him mosquito coils, which he was supposed to light and place under the bed at night. Not wanting to breathe the noxious fumes, Gary put it on the table next to the bed instead, hoping it would work.

Gary rented a car the next day, picked us up along with our friends Steve and Heather, and we had an enjoyable time driving around the beautiful island of Huahine. Gary is an avid fisherman, and arranged a trip with a local fisherman the next day, but when the fisherman's engine failed and the weather turned bad the trip had to be aborted. We had a farewell dinner on *Shaula* that evening, and said goodbye to Gary, who was departing on the ferry to Papeete the next morning. When Dan talked to his father on the ham radio a few days later we were shocked to learn that Gary had contracted dengue fever while at Huahine. This is a mosquito-transmitted disease, also known as break-bone fever because of the intense pain it causes in all of your joints. Gary felt ill and had chills and fever when he got off the plane in Los Angeles. The only treatment is painkillers, and it usually takes a couple of weeks to recover. We were really sorry his visit had ended this way.

Gary's visit had been enjoyable for us, but must have been somewhat of an ordeal for him. He endured lost luggage, an uncomfortable berth in our cockpit, a plugged up toilet, seasickness, and a cancelled fishing trip, but never complained. He must have wondered why we loved the cruising lifestyle, with all of its discomforts and almost none of the luxuries of land life. Although we love cruising on a small boat, we realize not everyone shares our enthusiasm.

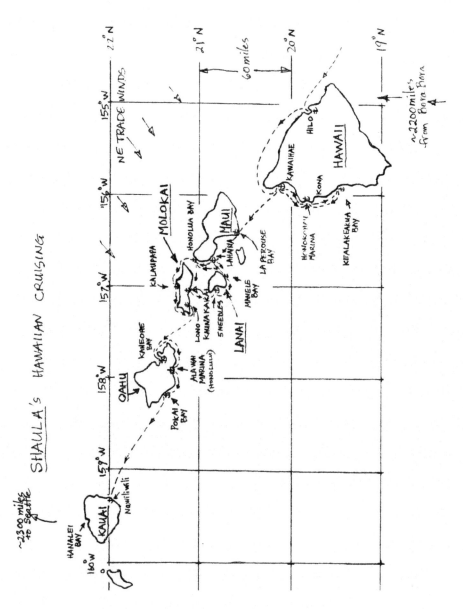

Cruising the Hawaiian Islands  November 1989 to June 1990

# 5 HAWAII

## Sailing against the Trade Winds

After our six-month stay in French Polynesia it was time to leave for Hawaii. It was mid-October and we wanted to be on our way north before hurricane season began in November. Our last stop in French Polynesia was Bora Bora. From there to Hilo, Hawaii is 2,220 miles. At our average speed of about 100 miles a day we expected the passage to take 22 days.

I had mixed feelings about the upcoming passage. It was sad to be leaving the beautiful and exotic French Polynesian islands to go to the familiar and much-visited Hawaiian Islands. Some of our cruising friends were continuing on west to the Cook Islands, Samoa, Fiji, or Tonga to end up in New Zealand or Australia for the hurricane season. We envied them. As usual I was apprehensive before leaving land to cross an ocean. Our last long passage from Mexico to the Marquesas had been wonderful because it was downwind sailing in the trade winds. Sailing from Bora Bora to Hawaii would be much more challenging. After crossing the equator we would have to sail north against the Northeast trade winds. The apparent wind is stronger when you sail against it, and the seas are worse when they hit your bow rather than coming from the stern. We didn't expect this passage to be easy.

As it happened, friends we had sailed with in Mexico and French Polynesia were leaving for Hilo about the same time as we were—Dave and Debbie on *Silverwind* and Steve and Heather on *Tandem Cay*. We were looking forward to visiting Hawaii together. We left with Dave and Debbie on the morning of October 17th. Steve and Heather left a day before us. As we left the lagoon in Bora Bora we took pictures of each other's boats sailing in the trade winds. It was a glorious start in beautiful sunny weather. Every day we all checked into the Maritime Net. In addition, we decided to have an informal net amongst our friends in the afternoons so we could find out how everyone was faring.

Unfortunately, the usual pattern of southeast trade winds south of the equator did not materialize—instead we had fairly strong northeast trades. Racing sailboats can sail within 45 degrees of the wind, but our sailboat is designed to sail downwind. The best we can do is about 60 degrees off the wind, and 70 degrees is more comfortable. This meant our course was northwest rather than northeast. Every day after we reported our position on the Maritime Net the Net Control would say "You'd better start getting your easting!" In the usual southeast trade winds this is good advice, because the strong northeast trade winds north of the equator can sweep your boat west past the big Island of Hawaii, so that you end up in Honolulu rather than Hilo. However, in the northeast winds we had, sailing northeast was impossible. It was very uncomfortable to be sailing so close to the wind, with the waves hitting the side of the boat and occasionally breaking into the cockpit. For the first eight days the motion of the boat was horrible—the boat jumping up and down and rocking sideways. To make matters worse the temperature was 85 degrees with 100 per cent humidity. It was difficult to cook, so we just ate sandwiches or salads. Dave and Debbie were having the same difficulties, with the additional problem that their lightweight boat pounded into the troughs of the waves when they sailed close to the wind. They had to reduce their speed to minimize the pounding.

Our daily informal net kept us sane through all our trials and adverse winds. It was an outlet for our frustrations in dealing with the discomforts of the passage. Nearly every day we and Dave and Debbie complained to each other about the awful conditions and our inability to head in the direction we needed to reach Hilo. However, Doyle, on the sailboat *Madame*, now in Hilo, came up on the net frequently to offer encouragement. He told us we wouldn't have any trouble getting our easting once we were north of the equator. It turned out he was right.

After our eight-day ordeal battling northeast winds we crossed the equator and had a wind shift to southeast winds (where we expected to have northeast winds). These southeast winds lasted for three days, and were a great relief for us—we could finally get some easting. Then the wind shifted, coming from all different directions—south, west, and southwest. We had frequent squalls, with winds of 25 to 30 knots, and some torrential downpours. This was the ITCZ, better known as the doldrums. This meant hard work for Dan, who was doing all the sail changes on our boat. He had to repeatedly reef and un-reef the mainsail, and take down or put up the jib and staysail. Although eight days of sailing against the northeast trades were terrible, this was far worse. The only compensation was that we could now easily sail on our desired course, northeast.

During this time we caught a nice tuna and a few days later a 43-inch wahoo. When Dan pulled in the wahoo we noticed a chunk of flesh was missing from one side. We could see circular tooth marks from the bites of a shark—this was probably a cookie-cutter shark. We had seen the fins of sharks following our boat previously after pulling in a fish, filleting it, and throwing the remainder overboard. This was the first time a shark had got to a fish on the hook before we pulled it in. We realized that sharks could be lurking anywhere. It was a reminder of why we never swam from the boat in the middle of the ocean, though we were tempted when we were becalmed on our trip from Hawaii to Seattle. We ate as much of the wahoo as we could, then cut the remainder into thin strips to dry in the sun for fish jerky. However, in the hot, humid weather and frequent rain squalls, the fish never dried out. It began to smell after a few days, so we tossed it overboard and the sharks got the rest of their meal.

Finally, on the 17$^{th}$ day of our passage, the strong northeast trade winds came in. We were 400 miles southeast of Hilo, and having made our easting we could lay the course. These were ideal sailing conditions, enough wind to push us along at over six knots on a beam reach (sailing with the wind blowing on the beam, one of most comfortable points of sail). It was blissful—a sleigh ride into Hilo. Our nightmare of a passage became a dream sail the last three days. We were delighted when we made it into Hilo in 20 days. Dave and Debbie had arrived that morning, but Steve and Heather still had a long way to go. They arrived six days later, exhausted by the effort of sailing closer to the wind than we, resulting in slower going, and continual bashing by waves breaking into the cockpit. They had taken seriously the advice to head northeast when they left Bora Bora, and had a much more uncomfortable passage.

What did we learn from this passage? The best course between two points on a small boat is usually not a straight line. The wind and seas determine the course, depending on how the boat performs. Our course from Bora Bora to Hawaii resembled a large question mark, starting with a straight stretch northwest, then sailing east, away from Hilo, then gradually circling west back towards Hilo. This made the trip easily doable in spite of adverse winds. Having daily contact with friends on other boats undergoing the same conditions made it much more tolerable—misery loves company. After making landfall it was a joy to get together with those who had made the same passage to celebrate and share experiences.

## The Delights of Hilo

It was time for some much needed rest and recreation. It was a luxury to be able to step off the boat onto the wharf rather than anchoring out and having to row ashore in the dinghy. Since our last visit four years before, new, improved showers had been installed in the restroom building at the marina, and next to it a large picnic table that cruisers could use for potlucks.

Radio Bay is a container port, where hundreds of containers are off-loaded from freighters day and night. Crane operators pick up and stack the containers in the yard just opposite the marina. The bright lights at night and the loud metallic clang of containers when they are dropped onto the pavement can be jarring and disruptive to sleep, although we soon got used to it. Radio Bay is also a terminal for cruise ships ferrying passengers around the five Hawaiian Islands. Cruise ships were arriving twice a week at the large wharf at the outside edge of the marina when we were there and could be noisy as they blasted their ship's horn on arrival early in the morning and before departure in the evening. However, on the day they were in port they provided a free van service every half hour from the marina area to the large shopping center, which we greatly appreciated. It was wonderful to be able to shop at the large supermarket and do our laundry for $2 a load, instead of the $16 it cost to wash a load in Bora Bora. At last we could buy all the foods that were unavailable in French Polynesia, such as whole-wheat flour, brown rice, celery, carrots, oranges and apples. We bought a cooler in which we could keep food on a block of ice, and replenish the ice as needed. On Saturday mornings we shared rides with other cruisers to the local farmer's market, where we bought fresh produce at a reasonable price.

What we liked most about Hilo was meeting sailors who had sailed to Hilo from all over the world. Sailboats arrived from Bora Bora, the

West Coast of the U.S., South America and Mexico during our stay. Everyone had endured a passage of several weeks, and we all enjoyed getting together to socialize and compare our experiences. Three days after our arrival we had a Mexican potluck at the picnic table at the marina with a dozen cruisers, including three boats that had arrived from Bora Bora, a couple who had arrived from California and an Australian couple with two young children. Every week or so new boats came into the harbor, and we had an excuse for another potluck in honor of the new arrivals. Steve and Heather on *Tandem Cay* arrived, and two boats arrived after long passages from San Diego. By Thanksgiving there were 14 sailboats in port, and we had a huge potluck on shore with turkey, dressing, cranberry sauce, potatoes, gravy, peas and carrots, and pumpkin and cherry pie. The feast lasted all afternoon and evening, and cruisers from five countries attended—Canada, the U.S., Peru, New Zealand, and Australia.

Hilo is not a major tourist destination, but has some nice, fairly inexpensive restaurants within walking distance of the harbor, where we could get a Chinese meal for $5 or a fish dinner for $8. A corner store, laundromat, a fish market, and a small shopping center with bookstores and a hardware store were also within walking distance. To visit some of the local attractions, such as Onekahakaha Beach State Park, the Hilo Tropical Gardens, and the Hilo campus of the University of Hawaii we took one- or two-hour walks. One morning we got up early to walk to the Suisan fish market auction held in a large building near Hilo Harbor. Arriving a half hour before the auction began, we watched the fishermen drive up in their trucks to deliver their catch. Dozens of huge frozen tuna were laid out on tables, each with a plug punched out of the side so buyers could see the quality of the flesh. The buyers sat in a small auditorium on the side of the room. Most of the buyers were from restaurants or large grocery chains, such as KTA and Sure Save. They bid on each fish with an imperceptible nod of the head as the auctioneer called out the bidding prices. We were astonished at the bidding price of some of the larger tuna—over $1,000!

Our friends, Dave and Debbie, went with us in a rental car to visit the Volcano National Park. We watched rivers of molten lava flow into the sea, and saw bursts of steam hissing as waves washed over the red glowing lava that had fallen from the cliffs.

So far we had avoided doing any serious boat work, but after a year of cruising the varnish was peeling off the exterior wood. In between all the great socializing, potlucks and sightseeing we worked on the boat. Dan sanded and re-varnished the cockpit combings, the main hatch, and the fore hatch. I cleaned the interior of the boat including the upholstery. Hilo, being on the windward side of the Big Island, is usually wet and windy, but when we were there in November and December 1989 they were having a drought. It made work on the boat, and life on the boat and ashore much easier. For a few days we had strong winds, which produced choppy seas, so we didn't leave the boat. When the wind died down we took a walk to nearby Reed's Bay and saw a 36-foot sailboat being towed off the rocks. Two multi-hulls also went on the rocks at Reed's Bay during the storm. We were glad we were in the safe harbor of Radio Bay, which is protected by a breakwater.

For Christmas Eve we and a dozen other cruisers were invited to a potluck by friends of one of the cruisers who lived a couple of miles north of Hilo. They had a spectacular view of the ocean from their home, and we had a wonderful dinner. Then on Christmas Day we had the largest potluck yet at the marina, with 25 cruisers. On Boxing Day we had another potluck of Christmas Day leftovers. Two days later our friends Les and Suzan on the boat *Sunrise* arrived from Mexico. Les, who was in his 40s, had suffered a heart attack when in Mexico, and had spent months recovering. It was great seeing him and Suzan, who were tired, but happy to have finally made it to Hilo. Dan rowed the quarantine and customs officers out to their anchored boat so they could check them in and also delivered a bottle of cold guava-orange juice, which they really appreciated. (We remembered how grateful we were when a fellow cruiser gave us a bottle of juice on our arrival at Hilo four years before.) And, of course, it was the occasion for another potluck the next day to celebrate their arrival.

Then the drought ended and we had several days of hard rain. It was time to leave Hilo and sail around to the warmer, dry side of the island, the Kona Coast. It had been a wonderful two months in Hilo, but we had to move on. The hurricane season begins in June in the Northern Hemisphere, so we had about five months to explore the rest of the Big Island, Maui, Lanai, Molokai, Oahu and Kauai. We looked forward to many more adventures before we had to head back to Seattle in June.

## The Dry Side of Hawaii

Sailing around the top of the island to the other side is less than 100 miles. Going from Hilo, on the windward side, to Kona, on the leeward side, is like going from Seattle to Los Angeles in terms of climate. On the Kona Coast we could do some necessary maintenance on the boat in the warm, dry weather.

In early January we sailed overnight to the marina at Honokahau. This marina was built by carving out a rectangular-shaped boat basin in the lava rock. With a narrow entrance it is well protected from the ocean swell. Charter fishing boats occupy most of the slips, but cruising sailboats are allowed to moor at the marina if there are empty slips.

When we arrived we were told we could stay for two weeks, at a cost of $79 per week—a very reasonable rate. The harbormaster, a big, macho Hawaiian man, allowed cruisers to extend their stay for a week or two at a time, but he favored women. So every couple of weeks I and women from other sailboats went to the harbormaster's office to ask for an extension, and we were always granted it. As this was the only marina on the leeward side of the island, we were thankful for his decision.

Dan resumed the re-finishing of our boat. He scraped the bits and the tiller and applied epoxy and varnish. This required a couple of weeks of dry weather. In addition the antifouling paint was flaking off. After applying antifouling paint every two years without removing the old paint we had built up a thick layer of paint on the bottom of the hull. Now it was flaking off in large patches because we had not adequately removed the wax on the hull eight years ago, when our boat was new. We decided to scrape off all the old antifouling paint down to bare fiberglass, and apply an epoxy barrier coat before painting on a new layer of paint. This would be a hard job, with lots of hand scraping, and I dreaded it.

Our friends Dave and Debbie, on *Silverwind*, had sailed to Honokahau a month before, and settled into jobs to earn money so they could continue cruising. They were living on their boat in the marina. Dave was working in the boat yard adjacent to the marina, while Debbie worked in a photography shop in the nearby shopping center. They needed to replace the foam on their V-berth and re-upholster all of the cushions on their boat. I had lots of sewing experience, and Dave had lots of experience scraping antifouling paint off boat hulls, so we reached a deal. I would help Debbie cut out and sew the upholstery on *Silverwind* while Dave helped Dan scrape the antifouling off the hull of *Shaula*. I helped Debbie shop for the foam, upholstery material, and zippers. Dave and Dan cut the foam to fit the sloping sides of the hull with a fillet knife. With Debbie helping in the mornings before she went to work at the photography shop, and on her days off work, it took us 10 days to re-do all the upholstery. I much preferred to be sewing on *Silverwind* than scraping and antifouling the hull of *Shaula*. With our boat hauled out in the yard where Dave was working, it took three long days for Dave and Dan to accomplish the huge job of scraping the hull, applying a barrier coat of epoxy, and re-painting. It was a fair exchange of labor and we were glad to have our hull well protected from barnacles and other marine growth. Dan still had to do the routine maintenance on the engine after 600 hours of use, and we had to replace our worn-out house batteries.

Dave and Debbie had bought a used car so she could drive to work, but they were happy to loan us their car if we would drop Debbie off at her job at the shopping center at Kailua-Kona. There we could buy our groceries at the big supermarkets, do laundry and re-supply with reading material at the Book Exchange. We enjoyed taking tours of the nearby businesses, such as the Kona Kai Coffee Company, where we tasted delicious coffee grown on the slopes of the Kona Coast, and Mrs. Fields Macadamia Nut Factory. We watched them crack the extremely tough nuts with a specially designed viselike nutcracker. Originally, macadamia nuts were used as jewelry, before anyone realized they were edible.

After we finished our boat work we decided to do some sightseeing beyond Kailua-Kona. Dan and I rented a car for $17 a day, although with the insurance, tax, and extra driver it came to $28 a day. The first day we visited Volcano Village, the Volcano Museum, and hiked through a lava tube. We then drove to Hilo to buy stove alcohol and granola. Driving back just before sunset we realized why the car rental place was called a Rent-a-Wreck. Besides having ineffective seatbelts, no emergency brake, and bare tires, the car had no visors. The late afternoon sun blinded us as we headed west, so we had to pull over to the side of the road and wait a half hour for the sun to set. The next day we drove to the Hyatt-Regency, where we were given a free one-hour tour of the facilities. It was the most expensive resort in the world. It cost $360 million to build, and had a huge swimming pool with slides and waterfalls, a Landing Bar, which you reached by electric boat through a narrow canal, and a dolphin pond, where guests had a chance to swim with the dolphins determined by lottery. Exotic birds sat on perches throughout the beautifully landscaped tropical gardens. It was a real fantasy land. Although we couldn't afford to stay in an expensive resort, we felt lucky we could visit this part of Hawaii in our own boat, where anchoring was free, and the moorage cost at the Honokahau Marina was less than $12 a day.

By the end of January many other sailboats had arrived at the Honakahau Marina from Hilo, and we enjoyed sharing rides into town, and having potlucks at the marina with many cruising friends. We all enjoyed walking to the small beach near the entrance to the marina, where we would spend an hour or so snorkeling in the warm water, seeing myriads of beautiful tropical fish. Other days we would walk over to the fish haulout area at the marina and see the enormous fish tourists had caught offshore—giant blue fin tuna, huge marlin, and a 1,200-pound mako shark. Sometimes fishermen gave us a marlin steak, or a small tuna, which provided some wonderful fish dinners.

When the winds calmed down, Dan and I sailed down the coast to Kealikekua Bay, a large bay about 10 miles south of Honokahau. A large monument dominates the shore, dedicated to Captain Cook, who was killed by the natives in this bay on his second voyage around the Pacific. Every day two large charter boats brought groups of tourists to see the famous bay and the wonderful tropical fish. The charter boats arrived in the morning and stayed for a few hours on a mooring buoy. Before they arrived each morning we moved off the mooring buoy and anchored in the more rolly part of the bay. When the last charter boat left in the afternoon we moved back onto their buoy so we could have a peaceful night without fear of dragging anchor. Every trip the tour operators fed the fish frozen peas, making the fish tame. Huge schools of all kinds of beautiful fish would come right up to our boat when we threw cracker crumbs into the water, and we loved to snorkel amongst them while they fed.

When we returned to the marina Dave and Debbie began a huge project of making a dodger over the companionway for their boat. Dave and Debbie had sailed across the ocean to French Polynesia and Hawaii without a dodger to protect them from sun, wind, and seas when sailing. Dave and Debbie wanted a dodger for their sail back to Takaroa in about six months, and eventually for their sail back to Canada. The $1,500 it would cost to have one made was prohibitive, so they planned to do it themselves. They bought the frame, material, zippers, and teak for the base. Dan helped Dave bolt the frame to the boat and build a teak arch over the sliding hatch to which the dodger would be attached. He and Dave designed the cover for the frame. Dave was determined to sew the many layers of material himself, but when I saw his wobbly first efforts at sewing, I volunteered to help. Other cruising friends loaned us their sewing machine, which we placed on a table on shore at the marina. With Dan and Dave helping to guide the material through the machine we managed to finish it in 10 days of hard work. Dan installed the snaps to attach it to the boat. It was tight fitting with zip-out windows, and looked almost professional. When we finished we were truly proud of the result, although we vowed we would never make another dodger. In appreciation for our efforts Dave and Debbie took us out to dinner at Harriman's, a very fancy restaurant owned by a famous chef who was a friend of theirs. It was a memorable end to our visit to the Kona Coast.

Our two-month stay on the Kona Coast was longer than we had planned, but we had a wonderful time in spite of all the boat work. The great camaraderie amongst the cruising sailors at the marina, and the sharing of work and play made it all worthwhile.

## Swimming with Dolphins

In Hawaii we were delighted to see pods of spinner dolphins. They're distinctive not only for their small, sleek bodies, but their habit of leaping straight up out of the water and spinning around on their axis for three or four turns before dropping back in the water in a belly flop. Usually, in a pod of a dozen or so dolphins most would come up slowly to the surface to breathe, while only one or two would do the spectacular spinning dance. We met a cruiser who told us he had the experience of his lifetime, swimming with the spinner dolphins. He was ecstatic. He said, "I've been swimming with spinner dolphins—it was a transformative experience. Look at me! Do I look different?"

Our boat was moored in the Manele Bay small boat harbor on the south shore of the island of Lanai. Locals call it Black Manele because the sand is made up of ground up black lava. Just a 15-minute walk along the road is Hulopo'e Marine Preserve. This bay, known as White Manele, is dotted with small coral heads and has a beautiful beach of white coral sand. The bay is open to ocean swell, so is not a good anchorage, but is a wonderful place to swim and observe colorful reef fish among the coral. We frequently put on our swimsuits and walked over to White Manele carrying our masks and snorkels to snorkel on the coral along the shore of the bay.

One afternoon we spent an hour or so snorkeling and identifying the many fish among the coral. Even in that warm tropical water we had to swim ashore and sit in the sun on the beach to warm up about every 20 minutes. Dan and I were both sunning on the beach when we spotted a large pod of spinner dolphins entering the bay. Dan was tired out after snorkeling, and I was still cold, but I didn't want to miss the chance to swim with the dolphins. I plunged back into the water and swam out to the middle of the bay where the dolphins were circling around in a large pod, coming up to the surface to breathe repeatedly. Swimming along the surface with my mask and snorkel I could keep up with their speed and observe them from above as they slowly glided along about 20 feet under the surface. They rose to the surface as a group, but I noticed that each time before they came up to breathe one dolphin dove deeper than the others and shot up to the surface vertically to spin around before falling back into the water. The next time they rose up to the surface to breathe I saw a small dolphin accelerating upward, so I lifted my head out of the water to watch, as the dolphin leaped clear out of the water and spun around several times. It was exciting to see them perform their spins up close. They put on this show repeatedly. Swimming with the spinner dolphins didn't really transform me into a different person, but it was a thrilling experience.

## Visiting Molokai

Molokai is an island just northwest of Maui. One of the cruising couples we met told us they had sailed to the north shore of Molokai, anchored in Kalaupapa Bay, and rowed ashore to join a tour of the nearby leper colony. This seemed like a great idea. We could avoid taking the steep trail down 1,700-foot cliffs on the back of a mule or on foot. Neither of us had experience horseback riding and hiking down and back up the steep trail would be hard on our knees, so we decided on the sailing option.

We left the anchorage at Honolua Bay on Maui in mid-morning on a beautiful, clear day with a warm 10-to-15-knot breeze behind us. Soon we were sailing along the north shore of Molokai, gazing at the most spectacular coast we had ever seen. Almost vertical cliffs rose from the clear blue water up into the sky. The cliffs were covered with dark green vegetation dotted with numerous waterfalls. With our self-steering vane keeping us on course we sailed downwind with main and jib in ideal sailing conditions, feasting our eyes on the fantastic scenery—it looked like photos we had seen in *National Geographic* magazines. We sailed for miles and miles beneath the sheer cliffs, awed by their beauty. It looked like an impenetrable fortress protecting the small leper colony nestled along the shore at the foot of the cliffs. The wind picked up in the afternoon, with gusts to 25 knots. As we rounded a long peninsula in late afternoon we were happy to arrive at Kalaupapa Bay, a large bay protected from the wind, but not from the constant swell that swept towards the shore. We dropped our anchor in sand at a depth of 30 feet, in the wind-shadow of the huge cliffs surrounding the leper colony. Large waves were breaking on shore, with spray blowing off the crests of the waves. By the time we anchored it was dinnertime, so we had a quick dinner of popcorn, canned lentil soup, and coleslaw. Dan put out our rocker stoppers, a stack of five plastic cones hung over the side of the boat from the end of our boom, held underwater by a lead weight on the end of the rope. When waves hit the boat rolling it from side to side, the rocker stoppers dampen the motion, changing it to a gentle roll, allowing us to sleep.

After breakfast the next morning we rowed in the dinghy towards the beach where we had seen a long barge-landing dock extending into the bay from shore. Unfortunately, we couldn't land our dinghy there because it was low tide and the barge dock was five feet above water level. To get ashore we would have to row our dinghy through the surf that was breaking on the rocky shore. Dan rowed the dinghy towards some large smooth boulders just above water level, and we managed to jump out between surges of the swell, one after the other, onto the rocks. Then we had to lift the dinghy out of the water before it got dashed on the rocks, and carry it up the boulder-strewn beach to put it in a safe place on the rocks above high tide. Carrying the 70-pound dinghy up 10 or 15 feet on a beach is easy, but carrying it over large boulders is another matter. With one of us on each side holding the rails of our dinghy we scrambled from boulder to boulder, trying to avoid falling and scraping our legs or the bottom of the dinghy on the rocks. This was the worst shore we had ever landed on with our dinghy. As we got closer to the settlement we heard a woman behind a house on shore shouting at us: "You can't come ashore without a permit!" By this time we were exhausted, having struggled for 20 minutes, our arms aching, to find a place where we could set the dinghy down. We walked over to the woman and explained that we had been told we could take a tour if we came ashore at the colony. She told us this was forbidden, but after watching our efforts to get ashore, she apparently didn't have the heart to turn us away. After a short discussion she relented, and signed us up to join the daily tour that started at 11 o'clock.

The woman invited us to rest on her veranda and served us some guava juice while we waited for the tour to start. Her name was Mary, the administrator for the colony. She was an attorney with a degree from University of Wisconsin. She was thinking of retiring when the job of administrator of Kalaupapa came up five years ago. When we visited, there were 94 patients remaining at the colony, with an average age of 68. Leprosy (also called Hansen's disease) is arrested by the use of sulfone drugs. If patients acquire immunity to sulfone they have to switch to other drugs, the newest one, at that time, having the side effect of turning their skin a deep brown, almost purple. Once treated their disease is no longer contagious.

The tours are run by one of the patients. They are called Damien Tours after Father Damien deVeuster, the priest who worked with the colony in the 1800s. Tourists are not allowed to take pictures of the patients. Each patient gets free treatment, housing, meals, and a spending allowance. Most food is brought in by plane, although a barge comes in with supplies once a year. It costs $3 million a year to run the facility. Hawaii feels an obligation to these patients who were sent here against their will to prevent the spread of leprosy.

As we spoke to Mary, we heard she thought the suggestion by a California senator that people with acquired immune deficiency syndrome be sent to such a colony to prevent the spread of AIDS was ridiculous. We sat there for an hour, fascinated by Mary's wealth of information. As the time for the tour approached, we realized that to join the tour we would have to carry our dinghy across the boulders again, row back to the boat to get money for the tour ($17.50 each), pack a lunch, then row back to shore through the treacherous surf, and carry the dinghy back up the boulder-strewn beach. Once was enough. We decided to forgo the tour. We thanked Mary profusely for allowing us to visit with her and apologized for intruding without a visitor's permit.

Not wanting to spend another night in the rolly anchorage of Kalaupapa we departed that afternoon and had another nice sail in moderate trade winds to the southwest corner of Molokai to anchor in Lono Harbor. It was protected by two breakwaters, and was very calm and quiet. We relaxed and reflected on what we had learned from our visit to Kalaupapa. Visiting the leper colony by sailing to Kalaupapa Bay and going ashore in a dinghy is not only very difficult due to the rocky coast, but is not allowed. However, sailing along the wild and beautiful north coast of Molokai was wonderful, and meeting the administrator and learning about the leper colony from her perspective was a unique experience. Though we missed out on the tour we were glad we had made the trip to Kalaupapa Bay.

## Sailing Home

Our two-year sailing adventure was sadly coming to an end. After seven months of cruising around the Hawaiian Islands it was time to return to reality and sail back to Seattle. Dan had a job waiting for him in September with the Seattle School District, and I hoped to find a job in Seattle. Cruising around the islands of Hawaii had been delightful and gave us a better idea of life in Hawaii than we had received from our short stays of two or three weeks in years past. However, most of the time we were at anchor or in a marina, with only short day sails between the islands. It had been seven months since we had made a long passage. The sail back to Seattle is 2,300 miles and meant a trip of at least three weeks, having to stand three-hour watches night and day. It would be difficult to adjust to the rigors of a long passage again.

It was mid-June by the time we were ready to leave. We provisioned with more than a month's worth of food, filled our water and fuel tanks, and each took a Bonine tablet just before leaving. The strategy for sailing to the West Coast from Hawaii is to sail north in the northeast trades until above the Pacific high. This adds many miles to the trip, but is faster than a rhumb line course. We were hoping the Pacific high would not be as extensive as five years before, making it a faster passage.

The northeast trade winds were strong when we left Hanalei Bay, on the island of Kauai, resulting in some boisterous sailing. Occasionally waves broke into the cockpit. The first day out I was thrown across the boat by a wave, catching myself with both arms against the cabin side. It felt as though my arms were dislocated at the shoulders, but the pain was only temporary. By the second day we adjusted to the motion, and regained our appetites, which are always poor at the beginning of a passage. However, it was difficult to cook because the alcohol we used to prime the kerosene burner kept sloshing out of the pan under the burner before we could light it. It sometimes took two or three tries to light the burner. Then I had to brace myself and open cans with one hand while I grabbed the rail with the other to keep from being tossed around the boat. When the seas are rough we had learned it was a good idea to cook a hot meal at lunchtime, because the seas were usually calmer earlier in the day than at dinnertime. After eating a good lunch we were able to just snack on crackers, fruit, or other handy food in the evening. The wonderful tropical fruit we had purchased at the farmer's market—mangoes, papaya, pineapple, and guavas—were a great treat. We were both eating and sleeping enough, enjoying the passage, and making fabulous progress north—130 to 140 miles a day. The weather was sunny and pleasantly warm, around 80 degrees, and we were down to shorts, T-shirts and bare feet.

Towards the end of the week the winds diminished, sometimes dropping to zero, so we motored for several hours each day. It was fairly comfortable in the flat seas. Pods of dolphins danced in the distance, and one pod of six large dolphins played at our bow for a half hour. One afternoon we saw a green glass Japanese fishing float bobbing in the swell in the distance. It was slightly smaller than the one we had collected on our trip from Hawaii to Seattle five years previously. This was an exciting find—most of the fishing floats we saw on this trip were plastic, and not worth picking up. We sailed over to it, went into irons to stop our motion, and were able to scoop it on board with our large fishing net. The bottom half of the glass float was covered with goose-neck barnacles, which Dan scrubbed off on the side deck.

The second week the weather became cold and overcast and the winds veered from northeast to northwest. The seas were confused and we had a very uncomfortable night. We skipped breakfast and lunch, but finally were able to cook dinner when the wind decreased. The next day we had no wind—we were in the Pacific high and had to motor for 11 hours. Fortunately, the high was a narrow band and we were able to pass through it in two days. For the rest of the second week we had fairly strong favorable northwest winds. It was foggy and drizzly, similar to Seattle weather. Now we dressed in layers—long johns, pants, long-sleeved shirts, jackets, wool socks and boots—and climbed into our sleeping bags at night for warmth. At times the fog was so thick we had to look around for ships every five minutes during our watches rather than the usual 10 minutes. It was scary booming along at six knots in fog and drizzle, unable to see more than a mile in any direction. If a ship suddenly loomed out of the fog in front of us, we would have to make a quick turn to avoid a collision.

During the second week Dan had the difficult job of replacing the badly chafed lines on our Aries self-steering wind vane. He had to hang off the end of the boat, detach the heavy wind vane and set it in the cockpit to accomplish the job. This required switching to the auto helm attached to the tiller for steering. A few days later he had to replace the blocks that the lines run through on the wind vane. From the first day of the passage we had noticed small, round, black objects falling into the cockpit well, but assumed they were papaya seeds. (We had been eating papayas in the cockpit). On closer examination we realized they were ball bearings that had fallen out of the blocks that the lines run through on the Aries wind vane. The blocks were warped from hard use. Dan had to replace the blocks with spares we had on board.

With the strong northwest winds we were making excellent progress.

We made our best daily distance of 160 miles. We were really flying. At the end of the second week we came out of the fog, but we had 90 to 100 percent cloud cover. We were certainly glad we had Sat Nav for navigation rather than relying on our sextant as we had on our Hawaii trip five years earlier. It would have been impossible to take a sun sight during this part of the passage.

When the winds strengthened to 25 knots in the wee hours of the morning I had to wake Dan to ask him to reef the main. Our starboard rail was dipping under the seas and the motion was horribly uncomfortable. It blew 16-to-20 knots all day. This was sailing at its worst. Even Dan admitted it was a bit grim. The only compensation was that at our fast speed we would be in port in another few days. However, three days later our luck ran out. The winds switched to southeast. A low pressure system had moved into our path. When we were 200 miles west of Cape Flattery at the entrance to the Strait of Juan de Fuca, we could make no progress to the east in the strong southeast winds. We had to heave to for a day, slowing the boat to about one knot. We spent the next three days with the wind and seas on our nose, sailing and motor-sailing at about half speed. Slowly the southeast winds changed to northwest, and we could sail without using the engine.

On the 21$^{st}$ day we began to see lots of kelp floating in large clumps, and a harbor seal popped up within five feet of our boat, sure signs that we were approaching land. At last, in the afternoon, we sighted the glorious dark peaks of Vancouver Island. Although we had seen only one freighter on the passage during the first 18 days, now freighters and fishing boats appeared all around us.

As we motored into the Strait the next morning we decided to skip Neah Bay and instead head straight for Port Angeles. The winds were finally favorable and we had three hours of pleasant sailing, arriving at the marina in late afternoon. We pulled into an empty slip and both of us got a full night's sleep for the first time in over three weeks.

All in all it was not a terrible passage. It had taken only 22 days, compared to our 28-day passage five years previously. Our previous experience helped. I was able to get enough sleep in spite of the motion. Dan helped with the cooking when the seas were so rough it was difficult to even light the burner on the stove. A hot meal during rough weather always lifted our spirits. Because the winds were light at times, and on our nose during the last part of the passage, we used the engine more than we liked. By the time we reached Port Angeles we had used a full tank of diesel (31 gallons) and were running on our 10 gallons of reserve that Dan had poured into the tank from jugs. But because we had provisioned for over a month's journey we didn't run out of food, and ate well most of the time.

After two years of cruising we were reluctant to leave our cozy home on the boat. Why should we return to Seattle so soon? It was only July 6$^{th}$, and school wouldn't start till September. With everything we needed to live on our boat except fresh food and Canadian charts, we decided to head for our favorite cruising grounds, Desolation Sound. We re-provisioned in Port Angeles, headed north to Victoria to buy some charts, and were on our way. The problems of buying a car and finding a place to live on land could be postponed for a month or so while we continued the life on the boat that we had come to love.

## How Our Relationship Survived

One of the most frequent questions people ask us is how our marriage survived a 15-year cruise on a small sailboat. Most people say they need time away from their partner. Their marriage would never survive such close quarters. During our travels we've met hundreds of couples sailing across the Pacific, some continuing on around the world. The vast majority are still together after years of cruising. It may be that those couples who take extended cruises are self-selected. Couples who require more space and time away from each other don't attempt to sail long distances together across the ocean.

For a dozen years before we began extended cruising we sailed in the summer from Seattle to Desolation Sound and beyond, first in our Rawson 30, and later in our present boat, *Shaula*, a 28-foot Bristol Channel Cutter. During the day-to-day sailing we didn't always agree on how to sail the boat. It was a matter of how close to the wind to sail, how to set the sails for best performance, and when to begin motoring when the wind became light. To save arguing, we eventually worked out a system where we took turns at the helm. Whoever was at the helm could sail the way they liked.

At the end of a day when we came into an anchorage we frequently argued over where to drop the anchor. Eventually we worked out a method that eliminates conflict. Dan takes the helm and motors around looking for the best spot to anchor, while I go up onto the foredeck and get the anchor ready to drop. When Dan finds the best spot he tells me the depth, and I let out enough chain to have a four- or five-to-one ratio of chain length to water depth. Dan then backs the boat to stretch out the chain, and we both watch the shore to determine that the boat has stopped moving and the anchor is set. If the spot Dan has chosen ends up too close to shore or to another boat, he has to pull up anchor and find a place to re-anchor. Because pulling up the anchor with 120 feet or more of chain with a manual windlass is a somewhat arduous task, Dan almost always picks a good spot the first time. When we leave the anchorage the next day, I pull up the anchor, because I enjoy the job and it provides good arm exercise.

During a passage it is easy to get along. With a three-hour on, three-hour off watch system the only time we talk much during the night is at the change of watch. The person off watch is usually sleeping or reading. During our watches we're both goal-oriented—trying to reach our destination as comfortably and safely as possible. Sometimes, when we're in heavy seas and I sight a ship, I'm unable to tell whether we're on a collision course. Our bow is zigzagging back and forth, making it difficult to determine the ship's bearing. I wake Dan for help in deciding if we need to change course. Dan goes back to sleep quickly, whereas the motion and the noise of water rushing past the hull keep me from falling asleep. When Dan woke me once from a sound sleep in the very early morning to inform me he had caught a large tuna I was furious. I value every minute of sleep I can get on a passage. Although I was thrilled with the tuna, I wished he had waited till my watch to tell me about it. He never woke me again during my off watch.

During the day we eat our meals during the change of watches, at 6 am, noon, and 6 pm. Dan checks into the maritime net during his off watch every morning, giving our position and weather conditions, and we listen to the reports of all the other boats making passages. If the wind picks up and we seem to be over-canvassed either Dan or I suggest it's time to reef. Dan is good about reefing when we first think it is necessary. Sometimes I call Dan to come up to the cockpit to see pods of porpoises, or some unusual birds, or a large ship.

Once we reach land we're no longer confined to our small boat. Our first job is checking into the country, which often takes hours. We always need to re-provision with fresh food, and we enjoy visiting friends on other boats. There are potlucks ashore, sharing music or meals with friends and sightseeing. The only time we feel cramped on board is in bad weather when we can't leave the boat. The floor space in the cabin is only nine feet long by three feet wide. However, we each have our own space in the main cabin, mine being the port side and Dan's the starboard side. We each have a settee with a reading lamp at its head, a bookshelf above, and a cabinet at the foot to store our belongings. This is where we sit, eat, read, and sleep. We may have cabin fever, but we're not in each other's way and there's not a conflict for space.

Serious disagreements were rare during our years of cruising. On our first passage when we reached Hilo I wanted to anchor in the inner harbor so that we could finally touch land after 24 days at sea. Dan preferred to anchor in the outer harbor and wait until morning to tie up to the wharf at Radio Bay. I finally gave up arguing and agreed to anchor. However, when I dropped the anchor it did not dig into the hard bottom, and our boat dragged across the bay. That ended the argument—I was greatly relieved when we motored into Radio Bay and tied ashore. The only other serious disagreement was about when to leave Cabo San Lucas on our first trip to Mexico. For the first 10 days we enjoyed our stay, visiting friends on several boats, having margaritas and whale watching at the Finisterre Hotel, eating at nearby restaurants, shopping for food, playing Pictionary with friends, and having potlucks on the beach. But then strong winds came in from the south, making it an exceedingly uncomfortable anchorage. Dan put out rocker stoppers to dampen the motion, but we were still rolling constantly side-to-side, making it difficult to sleep. It was also difficult if not dangerous to row the dinghy through the surf to land on the beach. One time a large wave turned the dinghy sideways to the swell. A breaking sea filled the dinghy with sand and water, soaking us to the skin. There was little to do when ashore except walk the beaches and shop for food. After a week of this I was ready to move on. A few days later Dan finally agreed to leave. He hadn't realized how tired I was of this anchorage.

Before we left for our first cruise I met a woman who had sailed around the world with her husband. As soon as they reached home port they got a divorce. During our cruising we met only a few couples who had serious conflicts. One couple was constantly bickering on land—we wondered how they survived being together on the boat. We knew another couple in which the husband set ridiculous rules on their small boat. He allowed his wife to bring only one shell aboard from the beach when she went beachcombing. He also rationed the matches his wife used to light the kerosene stove. In Fiji we witnessed a sailboat coming into anchor, the wife on the foredeck ready to drop the anchor, her husband yelling at her when she didn't drop it where he had asked her to. Another couple we knew had a yelling match down below when anchored in a small cove in Vanuatu. Everyone in the anchorage could hear them. We saw the woman dive off the boat and swim a long distance away. Then she circled around and swam quietly back to the boat but hung onto the stern, where she couldn't be seen by her partner from the cockpit of the boat. After a half hour or so her partner came out to the cockpit searching for her with his binoculars. We finally called him on the radio to tell him she was hanging on just behind the boat. We never learned what that argument was about.

It's hard to imagine what cruising would be like in a small boat if you were unable to get along. There is no way to escape. We were fortunate that we had very little conflict during our years of cruising. We've been married 48 years, still have our sailboat, and enjoy sailing together every summer in Desolation Sound.

## Moving Ashore

After returning from our two-year cruise we were able to sublease a slip at Shilshole Bay Marina. It was August 1$^{st}$, 1990, and the weather was beautiful. It was about 10 degrees cooler on the water than on land, so even on the hottest days we were comfortable. Since we had no place to live on land, we decided to continue to live on the boat. We had traveled about 15,000 miles and made three long ocean passages—from Mexico to French Polynesia, French Polynesia to Hawaii, and Hawaii to Seattle. During the passages, the boat seemed spacious. One of us was on a three-hour watch, with the run of the boat, looking around from the cockpit every 10 minutes, while the other was either asleep or doing other activities. When in port we spent much of our time out in the cockpit, or climbed into the dinghy to go ashore or visit people in nearby boats. Although the living space in our boat is tiny (eight feet by 16 feet), about the size of a walk-in closet, it seems completely adequate while cruising. Everything is compact, the main cabin serving as living room, dining room, kitchen and bedroom, with a bathroom in the forepeak.

When school started in September Dan resumed his teaching at Rainier Beach High School, and I began looking for a job. The weather turned wet and cool. We had no heater, so we wrapped up in sweaters or jackets in the cold mornings and evenings. It was a five-minute walk to use the marina restrooms or shower. On rainy days we both stayed below most of the day in the small cabin. On weekdays Dan had to leave early in the morning to drive to school. In the evenings he had to grade papers without a desk on which to spread out his work. Our small boat, that had seemed so comfortable while cruising, suddenly seemed cramped. We decided to find an apartment and move ashore.

In October Ethan, a math teacher at Rainier Beach High School, came to our rescue. Ethan was one of the crew members who had helped Dan sail our boat from Seattle to San Francisco on the first leg of our cruise. He had bought a house in Ballard and had just finished remodeling it. It was for sale, but until it sold Ethan offered to rent it to us for the cost of his mortgage payment, about $500 a month. When he gave us a tour of the house we couldn't resist. It was a luxurious home with three bedrooms, two baths, living room, dining room, large kitchen, and a full basement. It was like a mansion compared to our tiny boat. The next day Dan and I loaded our two boat mattresses and bedding into the car, packed up some clothes, food, and dishes, and drove to the house. We were looking forward to having central heat, hot and cold running water, a shower, an electric stove and refrigerator, a washer and dryer, television, and HiFi, luxuries we had missed during our two years on the boat. The evening we moved in it was pouring rain. It felt wonderful to be in a warm, dry place. We laid our mattresses out in the master bedroom and put on the sheets and blankets and pillows. The rest of the house was bare, but we planned to buy some used furniture to make the place livable. This was the first time in two years that we had slept anywhere except on our boat.

At 4:30 the next morning the slamming of a door awakened me. I assumed the man living next door had gone outside. A few minutes later I heard the distinct click of a doorknob being turned, sounding as though it was in our house. I immediately shook Dan awake and whispered, "There's someone in the house!" Dan jumped out of bed instantly and snapped on the overhead bedroom light. As he did so, a scruffy-looking young man in jeans and a red plaid shirt strode into our bedroom. I lay in bed and pulled the covers up to my neck, immobilized by fear. Dan, standing by the door in his boxer shorts and T-shirt, shouted angrily, "Who are you? What are you doing here?" The intruder replied, "I'm Frank Johnson, I own this house." Dan replied, "No you don't, I know the owner. You don't belong here. Get out!" The man just stood there, so Dan took him by the arm and marched him through the hallway and living room to the front door. Dan saw the man grab a burlap bag from the porch, then take off running up the street. Dan came back to bed, but needless to say we weren't able to get back to sleep. We wondered how the man got in and what his motive was. The house was empty of furnishings, so there was nothing to steal. Dan had noticed a strong smell of alcohol on the intruder's breath, so we guessed that he had spent the evening in a tavern. After much speculation on how and why this man had invaded our house, it occurred to us that we should call the police. Dan called 911 to report the incident, but by this time an hour had passed and the intruder was nowhere in sight. Two police officers came to talk to us, but since it was just a break and enter case, we suspected nothing would be done.

Before Dan left for work at 7 am, I told Dan I didn't feel safe in the house anymore. I wanted to return to the boat. Dan dropped me off at the marina before driving to work. I called Ethan to tell him what had happened. Ethan drove to the house later in the morning and discovered, from the muddy footprints on the driveway and on the back door, that the man had tried to kick in the basement door. He had gained entry by breaking a small basement window, climbing in, and walking up the basement stairs, and through the unlocked door at the top of the stairs. Ethan nailed a three-quarter-inch thick piece of plywood on the inside of the broken basement window, and installed a lock on the door at the top of the basement stairs. When we returned to the house that evening, we felt a lot safer. Ethan told us later that he had hired an ex-convict to help with construction work on the house. His name was Frank (although his last name wasn't Johnson). Ethan kept his toolbox full of his tools in the bedroom closet until the day before we moved in. He guessed that the young man had broken into the house to steal his expensive tools.

After returning to Seattle from our two-year cruise we suffered from cultural shock. There was traffic noise day and night. Instead of waking up when the sun rose, we now woke up to an alarm clock. Having to work five days a week and commute to work through heavy traffic was a shock. Although we enjoyed having a telephone instead of having to use pay phones, we were annoyed by frequent calls from people asking for donations or trying to sell us something. The biggest cultural shock of all was realizing that living ashore we could be victims of crime.

During two years of cruising through Mexico, French Polynesia, and Hawaii we never locked the boat at night. We felt perfectly safe. Only during our occasional stays at a marina did we lock the boat when we left to go ashore. No one ever broke in. To have a man break into our house on our first night ashore was more frightening than any experience we had on the boat. It seemed more dangerous to live on land than on our boat.

Arrival of our cruising boat from Costa Mesa, California

*Shaula* at anchor in Desolation Sound

Alice taking a sun sight with sextant on the way to Hawaii

Dan taking a noon sun sight on the way to Hawaii

Alice with Dan's parents, Helen and Paul, at the Kona Resort, Hawaii

Dan with tuna caught on the way from Hawaii to Seattle

Tom and Ethan on *Shaula* on the way to San Francisco

Dan pulling up rocker stoppers at Cabo San Lucas, Mexico

Alice whale-watching at the Hotel Finisterre, Cabo San Lucas, Mexico

*Silverwind* sailing along the Mexican coast

*Chautauqua* leaving Puerto Vallarta for French Polynesia

Dan taking down spinnaker using the sock

Dan with dorado caught on our trailing line

Dan with skipjack tuna caught on the way to French Polynesia

Alice in the cockpit with tuna strips drying on the boom gallows

Dan slicing my bread that failed to rise

Dan raising the French courtesy flag as we approach Hiva Oa

Dan and Alice hiking to the town of Atuona after anchoring at Hiva Oa

Cris and Dick playing bluegrass in *Shaula's* cockpit

*Shaula* at anchor at Hana Moe Noa, on the island of Tahuata

Dick rowing musical instruments ashore at Hana Moe Noa. Cris swimming behind the dinghy

Young girls at the Children's Dance Festival at Ua Pou

Dick on banjo, Cris on mandolin and Alice on fiddle playing bluegrass at Haka Hau, Ua Pou

Alfred bringing black pearl oysters to the dock at Takaroa

Hanere and Dave scrubbing marine growth off black pearl oysters

Ah Sahm drilling a hole for suspending a black pearl oyster in the lagoon at Takaroa

Nancy, Ah Sahm's granddaughter, enjoying a ripe coconut

Michael and Sheila's boat, *Kantala*, with local children aboard at Ahe, Tuamotus

Cris, Dick, and Alice playing bluegrass on the dock at Ahe, Tuamotus

*Chautauqua* and *Shaula* at anchor at Moorea, French Polynesia

Gary and Dan on *Shaula* sailing from Moorea to Huahine

*Shaula* at a mooring at Bora Bora

Alice and Debbie at the beach on Bora Bora

Women selling produce from roadside stand on Bora Bora

Dan with wahoo bitten by a shark caught in the ITCZ on the way from Bora Bora to Hilo

Dan working on *Shaula* in the boatyard at Hanakahau, Hawaii after removing all of the anti-fouling paint

Alice and Dan on *Silverwind* after completing Dave and Debbie's new dodger

Dan scrubbing off Japanese fishing float collected on the trip from Hawaii to Seattle

Shaula in a winter snowstorm in Seattle after returning from Hawaii

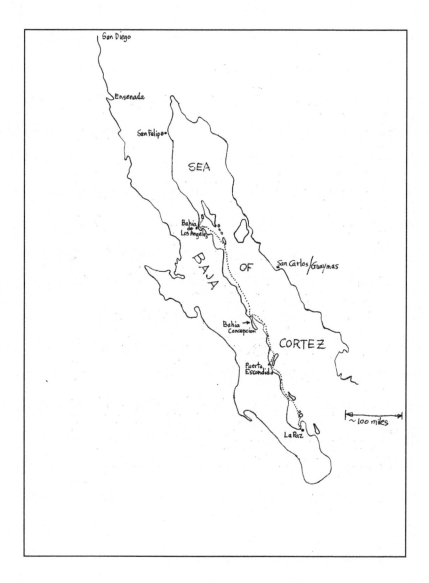

Cruising the Sea of Cortez  January 1994 to January 1995

# 6 SEA OF CORTEZ

## How We Could Afford Cruising

The whole experience of our two-year cruise was so amazing that we couldn't wait to go cruising again, hopefully for a longer time, and to more distant places. Due to the many good passages we had made I had largely overcome my fears of ocean sailing—fears of storms, and fears of hitting whales, containers or other objects. We planned to go back to work for as long as necessary so that we could quit our jobs and cruise indefinitely. Now we had a huge incentive to save money—the more we saved the sooner we could return to the cruising life that we loved.

When cruising for two years our expenses averaged $1,000 a month. There were times when our expenses were much less, especially when crossing the ocean, where there is no way to spend money. In Mexico food was cheap and there were few marinas, saving us the expense of marina fees. We anchored everywhere in French Polynesia, and only stayed for short times in marinas in Hawaii.

In 1992, two years after returning to Seattle from Hawaii, Dan got a lucky break. That year, in order to save money, the State of Washington offered early retirement. State employees could retire and start collecting retirement at age 55 instead of age 60 if they had at least 15 years of service. This was perfect for our plans to go cruising. Dan was 56, and could retire at the end of the school year in June. During the next year I continued to work at the Blood Center while Dan painted and varnished the interior of the boat, installed larger solar panels and a watermaker, updated our electronic equipment, and did numerous other jobs to get the boat ready for an extended cruise. Now we had a goal of leaving in the summer of 1993 to cruise for as long as we still enjoyed it.

In June 1993 I quit my job, we sold our car and our furniture, put our other belongings into storage, and moved onto the boat. Dan's retirement check was enough to pay for most of our expenses on our cruise, and we supplemented it by withdrawing $500/month from his 403b retirement account. We would be able to live comfortably on Dan's $11,000 a year retirement plus the $6,000 per year from the 403b. Four years into our cruise I could start collecting Social Security at age 62. This meant we would no longer have to take money out of the 403b. It was not sufficient income for living on land, but was enough to live well on the boat.

## The Lower Sea of Cortez

The Sea of Cortez is a long, narrow stretch of sea between the 775-mile long Baja peninsula and the Mexican mainland. Hundreds of sailboats cruise in the Sea of Cortez, mostly from California, but also some from Portland, Seattle or British Columbia. The Sea is protected from the ocean swell, and teems with all kinds of sea life: skates, whales, seals, sea lions, and numerous fish. There are many coves for anchoring both on the peninsula and on small islands.

When we made our first cruise from Seattle down the coast to Mexico and to French Polynesia we were on a schedule. We had to be back in Seattle at the end of two years so Dan could resume his teaching job. We skipped the Sea of Cortez and headed down the mainland coast of Mexico to Puerto Vallarta in order to leave for French Polynesia in the spring. Now that we were retired and had no schedule we had the luxury of being able to cruise the Sea of Cortez for a year and leave for French Polynesia the following year.

Hurricane season in Mexico extends from May until November. This means cruisers usually arrive in Mexico in November and leave for French Polynesia, Hawaii, or the U.S. West Coast in April or May. After sailing to Mexico in November 1993, we entered the Sea of Cortez in January 1994 planning to stay a full year. It is considered safe to cruise in the northern part of the Sea during the hurricane season because hurricanes are very rare there. We committed to staying in the northern Sea of Cortez at least until the end of the hurricane season in November.

For the first two months we hung out in La Paz and the many nearby anchorages. In March we sailed north to Puerto Escondido, a large harbor about a third of the way up into the Sea. Puerto Escondido is a large bay with room for dozens of boats. Cruisers often gathered on shore to walk or drive to the small community of Tripui for fresh produce. Once a week we gathered for Spanish lessons on the patio of a Tripui restaurant.

The seas are generally flat except in strong winds. Sometimes we caught fish using a lure dragging behind the boat, but mostly, after anchoring, we fished from the dinghy using fishing poles. It was as easy to get fresh fish for dinner as shopping at a fish store, except we couldn't choose the kind of fish. That didn't matter; they were all delicious—sierras, mahi-mahi, Bonita, black skipjack, jack crevalle, yellow tail, and triggerfish. Whenever we caught a large fish, or several smaller fish we would invite friends over to our boat for dinner. Everyone we knew would reciprocate when they caught excess fish.

One afternoon when we were on the boat we heard beautiful flute tones coming from one of our neighbors. I got in the dinghy and rowed around to find the source of this pleasant music wafting across the water. To my surprise I discovered it was from the smallest boat we had seen in Mexico, *Scout*, a 25-foot Pacific Seacraft sailboat. A couple in their 30s, Lynn and Rob, invited me to come aboard. Lynn was learning to play the flute with the help of Rob, a jazz musician who played the trumpet. He had taught Lynn how to read music, and she had worked her way through a flute exercise book and was quite proficient. She accepted my invitation to come over to our boat to play flute-violin duets the next day, and this became a regular afternoon entertainment. We had many afternoon practice sessions over the next five or six months, playing either my music for two violins or Lynn's music for two flutes. On another boat in the anchorage, *Juniata*, we found another flute player, John. He was actually a bassoon player, but because a bassoon was too large an instrument to take aboard a boat, he was learning to play the flute. He had trio music for three flutes, so he would often join Lynn and me on our boat. We also met guitar and keyboard players. I enjoyed playing violin with various combinations of instruments.

At many potlucks on shore we met a few cruisers who were planning to sail to French Polynesia in the spring, and we shared our experiences of crossing the Pacific and visiting these islands that we had enjoyed so much. We had been reading about 1995 being an El Nino year, and we were somewhat concerned that this would not be a good year to cross the ocean. We had many discussions of the pros and cons, but in the end decided that we would take our chances of leaving the next spring. It was an unknown risk.

## A Close Encounter

While anchored at Puerto Escondido we took bird walks to Rattlesnake Beach along a dirt road, sometimes with bird-watching friends from other boats. The Baja Peninsula is like a desert, with very little vegetation, except for many kinds of cactus. The birds we saw were flitting among the small shrubs and cacti. We saw beautiful yellow orioles, gnatcatchers, kestrels, ground doves, gold finches, hummingbirds, and cactus wrens.

One afternoon Dan and I took a walk along the dirt road to Rattlesnake Beach. We turned off onto a side road and nearly stepped on a large rattlesnake lying across our path. It was fat and lethargic, and did not move out of our way. Not daring to disturb it by stepping over it, we backtracked and continued down the main road. After walking along the road for 20 minutes, seeing only a handful of birds, we glimpsed a small gnatcatcher in the trees near by. Dan made his high squeaking sound by sucking on his closed fist, imitating a bird in distress. His calls are usually effective in enticing small birds to come closer to check out the bird they hear. After several minutes with no bird appearing we noticed a large animal creeping stealthily out from under the brush along the road 50 feet ahead. It was a large bobcat. Apparently Dan's birdcall was effective enough to convince the bobcat there was a bird in distress in our vicinity. We watched as it turned towards us and began to stalk us. We decided to abandon our bird walk, and quickly turned around to head back to the harbor. Afraid that if we began to run the bobcat would give chase, we sauntered back down the road, with the bobcat following. Every time we glanced back there it was, keeping up with our pace, although not gaining on us. It was unnerving, to say the least. What if after risking our lives sailing in big seas and strong winds to get here we met our end in the desert being mauled by a bobcat? Finally, after 10 minutes, we noticed that it was lagging behind, apparently no longer convinced there was a bird with us. We continued back to our boat. It was the scariest bird walk we had ever taken. Instead of reporting the many birds we had seen we amazed our friends with our story of being stalked by a bobcat.

## Chubascos

Before we sailed into the Sea of Cortez, I thought chubasco might be a spicy dish served in a Mexican restaurant. While in the Sea of Cortez we frequently checked into the Chubasco Net in the morning. This didn't sound sinister; I thought it was just a pretty name for a cruisers' net. From reading stories of round-the-world cruisers I had a vague impression that a chubasco was a strong local wind found along the coast of South or Central America. We were soon to learn the meaning of a chubasco first hand.

In mid-June we were peacefully anchored at San Marcos, about half way up in the Sea of Cortez. We had daytime temperatures of 96 degrees, but the water was 76 degrees, so it was possible to cool off by swimming. Our large shade awning, which we rigged above the boom from the mast back to the end of the cockpit, kept life tolerable on the boat in the heat of the day. While at anchor we left it up day and night. The fourth day at anchor at San Marcos we woke up to a heavy downpour, thunder, and lightning. It was the first big rainstorm in three months. It was cloudy all day, and we relaxed on the boat. When the rain stopped we rowed around the anchorage to visit the cruisers on three other boats anchored there. That evening at 11 pm we were wakened by a sudden strong wind of 20-to-25 knots, with one huge gust of 38 knots. The boat ahead of us, *Mizpah*, let out 60 feet of anchor line, putting their boat within a boat length of ours. To keep their boat from running into ours we let out more chain, for a total of 225 feet. With a ratio of 10 to 1 (the ratio of length of chain to depth of the water) we were unlikely to drag anchor. Next we had to roll up the big awning, which was flapping wildly in the gusty wind. This is not an easy job in the strong wind, as we had to untie the lines at the ends of the poles where they were attached to the lifelines, and detach the lines from the back stay in the stern of the boat. Then with one of us on either side of the boat we rolled the awning forward, wrapping it and the poles in a tight roll, and tying it down in a large bundle on the front of the boom. With the awning taken care of we pulled the dinghy, which was dancing wildly at the stern, alongside our boat. We lifted it the usual way by cranking it up with the main halyard winch, then turning the dinghy upside down, and lowering it onto the cabin top. In the strong winds these jobs were extremely difficult. At last we went below and tried to sleep as the wind howled in the rigging and the boat bounced around in the choppy seas. At 3 am, as suddenly as the wind came up, it died out. Sleep at last! That was our first taste of a chubasco.

From San Marcos we harbor-hopped up into the northern Sea of Cortez. The sails between the anchorages were delightful in the moderate winds and flat seas. The daytime temperatures were usually in the 90s, cooling down to the 80s at night. The water was now about 90 degrees, so jumping in the water did little to cool us off—it just made us stickier. Being close to Arizona we expected to have dry heat, which is more tolerable than when the humidity is high. However, it was hot and humid in the northern Sea of Cortez because the prevailing breezes were from the tropical South. It was uncomfortable in the day, and difficult to sleep at night in the sweltering heat.

One afternoon in mid-July we sailed from our anchorage on Isla Smith over to the village on the Baja Peninsula to buy fresh food. While shopping we ran into our friends Gail and Bob from *Tulum III* and Craig from *Maxwell's Demon*, and had a great visit, chatting and eating ice cream cones till 7 pm. After returning to the boat we made fish tacos from a black skipjack we had caught on our trolling line while sailing to the village. That night at 11 pm we were again hit by a sudden blast of wind from the north. Earlier in the evening I had seen lots of lightning on the mainland east of us, but never suspected we would be hit by another chubasco. Fortunately we had not re-rigged the awning after our sail to the village, but we had to go up on deck in the howling wind and haul the dinghy out of the water to put it on the cabin top. We saw gusts of 30 knots on the anemometer. Then the wind picked up from the east at 20-to-25 knots. By midnight it had switched to the south, and we had beam seas, which meant the boat rolled uncomfortably from side to side. Although we were unable to sleep, we had no fear of dragging, because we had 170 feet of chain out in 22 feet of depth, a ratio of 8 to 1. We stayed awake till 2 am when the wind finally died down.

Ten days later, when we were anchored at Isla Smith, we were woken at 3 am by a huge gust of wind from the south. At 45 knots it was the strongest wind we had ever experienced. We quickly closed the fore hatch and skylight hatch and brought in our towels and swim suits from the lifelines. During a 20-knot wind earlier in the evening we had already rolled up the awning, but once again we had to haul the dinghy onto the cabin top. There were huge black clouds overhead, and we had lots of thunder and lightning. The wind finally decreased an hour and a half later, replaced by heavy rainfall. At last we could go back to sleep.

During August there were two more chubascos, taking us by complete surprise, with huge gusts of hot wind hitting the boat like a blast furnace in the wee hours of the morning. Quickly jumping out of bed and scrambling out on deck to roll up the awning and raise up the dinghy in horrible conditions was getting tiresome. We lost many hours of sleep while the boat bounced around, lightning flashed, and the wind howled through the rigging. Fortunately, we never sustained any damage to our boat or dinghy. Other cruisers fared far worse. In the last chubasco in mid-August two boats dragged anchor at the anchorage at Don Juan, *Escapade*'s awning ripped in two places, and *Dream Catcher*'s dinghy flew horizontally from the stern of their boat. They had to cut the line and let the dinghy blow ashore. They retrieved the dinghy from the beach the next morning. The following day on the Chubasco Net two boats reported that their dinghies had flipped over and they lost everything in them, including the oars. One couple had left their outboard on the dinghy, which had flipped over submerging the engine. They were unable to start the outboard the next morning.

## Northern Sea of Cortez

During the next four months we sailed further north, anchoring at small bays and the many islands near the Baja Peninsula. In August we sailed to Puerto Refugio, on a small island about two-thirds of the way up in the Sea of Cortez. It was too hot and humid to take walks on shore except in the early morning before the sun came up. Instead we jumped in the water to swim or snorkel to cool off. Several of our friends on boats anchored in the same bay decided they wanted to swim with the sea lions. We had seen a large colony of sea lions gathered on the beach of a small island we had passed, Isla Granita. Our friends with large dinghies invited us to join them for the three-mile trip to the island. Eleven cruisers piled into several dinghies. Everyone was dressed in swimsuits and Lycra suits. The lycra suits zip up to the neck and completely cover the body to protect swimmers from jelly fish stinging cells that are everywhere in the Sea of Cortez. We carried our masks and snorkels and motored over to the island. We anchored the dinghies, jumped overboard, and began swimming back and forth near the beach, observing the sea lions swimming in the fairly shallow water adjacent to the island. Some large sea lions remained on the beach and began barking and making loud deep-throated trumpeting sounds. They seemed to be warning us off—perhaps they were defending their territory.

It was an amazing experience swimming with sea lions and observing their graceful and effortless glides and turns. However, while I was floating at the surface observing a small sea lion as it swam along the bottom, it suddenly came up to the surface right beside me, with a menacing look. I got the impression that he did not like us swimming in his waters. It's hard to imagine what sea lions are thinking when they see these long, thin mammals in their brightly colored Lycra suits and masks and flippers, but they could see us as rivals, taking the fish they feed on. Several of our friends had their spears with them, and caught a grouper and a sheepshead in this area. After a half hour of snorkeling along the island shore, listening to the constant din of the barking sea lions, we climbed back in the dinghies and departed back to our boats.

We had all enjoyed the afternoon, but wondered if it was a good idea to invade the territory of these wild creatures. Just as a bobcat is foreign to us and unpredictable in its intentions, sea lions are an unknown. Perhaps we were foolish to swim amongst them—they could easily bite an arm or leg before we realized they were hostile. If I had it to do over again, I don't think I would chance swimming amongst a colony of sea lions. With 50 or more sea lions among our small group of cruisers we were far outnumbered. They also far outweighed us and were much more swift and agile in the water than we were. Luckily, we all came out unscathed.

In July we sailed further up the Sea to Isla Smith, another favorite anchorage for cruisers. We were delighted to find Rob and Lynn in *Scout* anchored nearby. We had first met them in Puerto Escondido. Lynn and I resumed our duet playing in the afternoons. While there were many activities we enjoyed while at anchor in the Sea of Cortez, such as fishing, swimming, snorkeling, beachcombing, sharing meals, happy hours, and potlucks, playing music was a high priority for me. It is always a delightful surprise to find musicians on other boats. It's much more enjoyable to play music with friends than to just practice alone. Lynn and Rob were planning to sail to French Polynesia the following year, as we were, and we hoped to get together in Tahiti to play more duets

In August, it became very uncomfortable. With temperatures up to 100 degrees, and high humidity, we sweltered day after day. There was very little breeze, and often flies swarmed around the boat. Normally we could jump in the water and swim around the boat to cool off, but in the northern part of the Sea the water was 95 degrees, and didn't provide any relief from the heat. Once there we stayed, however, because it's not safe to leave the boat unattended in case of a hurricane.

To escape the heat we could have sailed over to San Carlos on mainland Mexico, left the boat in the marina, and traveled to Southern California. Instead, we waited until October before sailing south to San Carlos. We left the boat there for three weeks. After eating lunch in a small Mexican restaurant in San Carlos we boarded a Mexican bus headed for San Diego. Unfortunately, the air conditioning system on the bus had broken down in that it couldn't be turned off. I had a bad cough, and Dan had Montezuma's revenge from the meal at the restaurant. It was a miserable trip. When the bus stopped in Tijuana Dan called his brother, Gary, who suggested we take the Tijuana trolley to San Diego. Gary picked us up at the train station, took us to his home in Oceanside, and fed us chicken soup. After several days of recuperation we were able to visit Dan's parents in Glendale.

In January we left the Sea of Cortez, sailed down the mainland for a month or so, then to Puerto Vallarta, the jumping-off point for French Polynesia. This time of the year it was much cooler and more pleasant than it had been in mid-summer in the Sea of Cortez.

The Sea of Cortez was not the paradise we had imagined. The high summer temperatures and humidity were almost unbearable, and the chubascos were unpredictable and un-nerving. It is good that we didn't know about the summer heat and chubascos before we sailed there. If we had been forewarned we might have sailed on to French Polynesia and missed the great camaraderie among fellow cruisers, the frequent potlucks ashore, and the excellent fishing in the area.

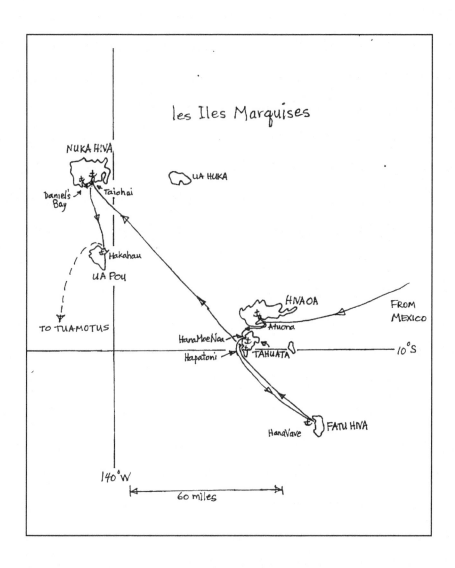

*Shaula's* passage from Mexico to the Marquesas

## SAILING THE SOUTH SEAS

Shaula's passage from the Marquesas to New Zealand

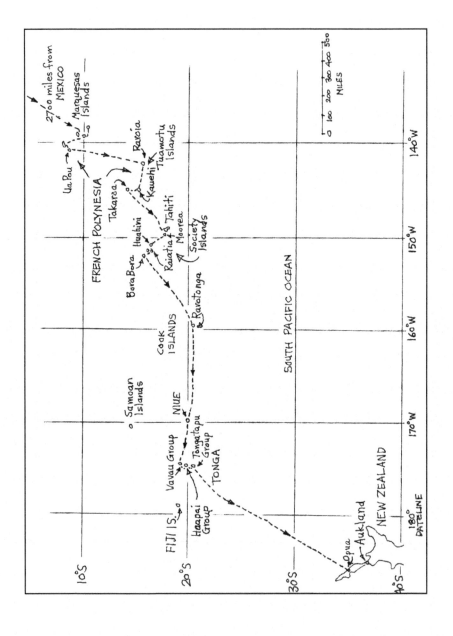

# 7 MEXICO TO NEW ZEALAND

## Sailing to the South Pacific

In March 1995 we left Puerto Vallarta for the 2,700-mile voyage to the Marquesan island of Hiva Oa in French Polynesia. When we made this same passage to French Polynesia six years earlier we found the sailing to be wonderful. This is the longest distance of any passages we planned to make. From French Polynesia we would be island hopping to Rarotonga, Niue, and Tonga on the way to New Zealand, all passages of less than 1,000 miles.

Our boat was better equipped than on our first trip. We had a new set of sails, and Dan had installed roller furling on the jib to make it easier to increase or reduce sail. This was a huge improvement, as we could now furl and unfurl the jib without leaving the cockpit. Previously, Dan had to climb out on our narrow seven-foot long bowsprit to pull down the jib and secure it. He sometimes was doused to the waist in saltwater in the process. We had a new dodger made, replacing a narrow dodger we had installed for our first trip. It provided us much better protection from wind and seas when in the cockpit. We had a GPS for satellite navigation, superior to our previous Sat Nav, and much easier than taking sun sights with a sextant.

After days of shopping for provisions, topping up our water and diesel tanks, and filling a couple of five-gallon jugs with extra diesel we were ready to depart. Dan lashed everything on deck so nothing would be washed overboard, while I stowed everything down below so nothing would fly around in the cabin. We put on our harnesses and hooked our tethers to the eyebolt in the cockpit, and we were off.

As soon as we were underway we set up a three hours on/ three hours off watch schedule. Other cruisers we know keep longer watches, four or six hours, and some women keep an eight-hour watch at night so their partners can sleep. Neither of us wanted to stand watch for more than three hours at a time so this schedule worked for us. The most important duty on watch is to look out for ships. That means looking around 365 degrees every 10 minutes. Climbing the ladder six times an hour from the cabin to the cockpit to look around provides our exercise and keeps us fit. If we see a ship we check its bearing. If the bearing doesn't change, it means we're on a collision course. Even though a boat under sail has the right of way over a powerboat, we don't expect a large ship to change course to avoid us. Large freighters can only turn slowly, whereas we can push the tiller over and change course in an instant. On this passage the sighting of ships was a rare event. During the first week after leaving Mexico we saw three freighters and the loom of a fishing boat, but all crossed our course a long distance away, so there were no near collisions. It was another 10 days before we saw another freighter. It was still necessary to keep a watch—ships could appear anywhere, anytime.

The person on watch is the captain and has complete autonomy. He or she can change course, change the set of the sails, or choose to motor if the wind becomes light, while the off-watch person goes below to read, relax, or sleep. If the wind shifts, the self-steering wind vane changes our course, and we have to change the angle of the wind vane to bring the boat back on course. Every hour we write down in our log our position, speed and compass course, the wind speed and direction, and the miles to go to our destination. This gives the person coming on watch a quick summary of what's happened while they were down below.

During our watches we enjoy watching the rolling waves progress across the deep blue ocean, and the fluffy trade wind cumulus clouds gliding across the sky. Our boat is in the center of a six-mile diameter circle that drifts across the ocean at a speed of about five knots. Within this circle we have a 365-degree ocean view. We have spectacular views of sunrises and sunsets, and at night a bright moon and a multitude of stars. The days are filled with delightful sightings of birds and sea life. Several times on this trip we saw large pods of small dolphins leaping out of the water, and schools of flying fish soaring as they were chased by larger fish. Occasionally in the mornings we found dead flying fish and squid that had landed on our decks at night. One evening when I discarded food waste overboard I saw a shark swimming in our wake. Small flocks of storm petrels and shearwaters soared over the tops of waves, and rarely a beautiful red-tailed tropicbird flew high overhead. There were nuisance birds—ungainly, stupid boobies that tried repeatedly to land on our boat for a free ride. If allowed to stay on the boat they could leave a terrible mess. A couple of immature boobies tried without success to land on our boom, and a red-footed booby landed on our lifelines near the bow of the boat. Dan went forward on the side deck waving and shouting, and got within three feet of the bird before it flew off. When a booby landed on the base of our wind vane, it refused to leave until Dan swung a cockpit cushion at it, grazing it on the wing.

The motion of the boat was a comfortable rise and fall on the widely spaced rolling waves. I had learned from our previous passages that the motion often becomes wild around dinnertime, so on this trip I made a hot meal at noon. Even though we had no refrigeration we ate well on a combination of canned food and fresh vegetables—such as stew made of canned beef with tomatoes, carrots and potatoes, garbanzo spaghetti, creamed chicken on noodles, bean tacos, vegetarian frittatas, or spaghetti with tuna sauce. For breakfast we usually ate granola and fruit, although several times I made pancakes or scrambled eggs with hash browns. At dinnertime we snacked on whatever we wanted: sandwiches, crackers, corn chips, dried fruit, cookies, or chocolate bars. One morning I managed to bake two loaves of bread and two chocolate cakes. Unfortunately, the gimbaled stove pitched forward when I opened the oven door, and one of the chocolate cakes fell onto the cabin sole. I was able to scoop it back into the pan and we ate it from bowls using spoons. I called it chocolate crumb cake. Dan always did the dishes at the end of the day, which I greatly appreciated. Towards the end of our passage we caught a nice size mahi-mahi within three hours of putting out the trolling line. It made a wonderful lunch of mahi-mahi almandine, and a dinner of teriyaki mahi-mahi.

The winds were light to moderate for our whole passage, with no gale force winds. Dan was kept busy reefing the main and rolling in the jib when the winds picked up, and unrolling the jib and unreefing the main when the winds lightened. During the first week, Dan frequently set the spinnaker, which we would fly for hours and even overnight when the winds were too light for our working sails. When the wind died completely we motored to keep our speed up and to charge the batteries. While motoring our electric auto helm attached to the tiller steered the boat. All we had to do was push buttons to correct the course, either one degree or 10 degrees left or right. It kept us on a much straighter course than we could steer by hand.

Just north of the equator we passed through the ITCZ, an area that can produce calms, known as the doldrums, and squalls with heavy rain. We were fortunate that the ITCZ at that time was a narrow band which we managed to get through in a day by motoring directly south. We only experienced mild squalls and a few rain showers. After we crossed the equator we were in the southeast trades and had some wonderful downwind sailing. By frequently changing sails to suit the conditions we made fabulous progress, averaging over 120 miles a day. Our best 24-hour run was 184 miles, helped by a two- or three-knot favorable current.

On Easter morning we were excited to see the island Hiva Oa after only 22 days at sea. I was enjoying the passage so much that I didn't want it to end. However, the sight of mountainous island covered with lush green jungle and the smell of land was inviting. We entered the small harbor at Atuona, anchored among 10 other sailboats, and celebrated the best passage we had ever made.

## Communicating on a Passage

Sailors who sail across the Pacific Ocean and have an amateur radio license can check into the Pacific Maritime Net. The net controllers are dedicated ham radio operators who keep track of cruisers during their passages. As each boat checks in at the beginning of their cruise, the net control takes down their particulars, such as boat name, size, crew members and destination. The boat is placed at the bottom of the roll call list. Once a cruiser checks in they are committed to respond every night when their name is called to report their position, speed, course, and wind speed and direction. Each evening before roll call the net control asks if there is any medical, emergency, or priority traffic. Every evening we heard reports from other boats in our vicinity, and were forewarned of any bad weather or calms. We plotted the positions of our friends to compare their progress with ours.

At the end of a boat's report, if the propagation is good the net control may ask whether the person wants to talk to anyone on land. If so, they set up a phone patch. They shift to a different frequency, call the person on land, and allow both parties to talk for free. The only difference from a regular phone conversation is that each party must say "over" when they finish talking so the net control can switch his radio back and forth between the two parties. There was a net controller who lived in Seattle who set up phone patches so we could talk to friends and relatives there, and another in Southern California who set us up to talk to Dan's parents in Glendale. We were able to get phone patches frequently. It was a great way to learn the latest news from home, and reassure friends and relatives that we were all right.

Before leaving for our second passage from Puerto Vallarta to the Marquesas we met with a group of people who were planning to leave for French Polynesia. For most people it was their first long passage and they wanted to hear about the passage we had made six years before to allay their fears. At the meeting we decided to set up an informal SSB net that could include non-hams for those of us leaving for the Marquesas in late May, and called it the Late-Leavers Net. It started out with only three or four boats, but built up to 20 boats during our passage. Most people checked into the Westbound Net in the morning and the less formal Late-Leavers Net in the afternoon to discuss problems during their cruise. Dan checked into the Pacific Maritime Net in the evenings in addition. This took two to three hours a day, but it provided useful information and camaraderie with other cruisers. We arranged our watch schedule so Dan would be off watch during those times.

Our passage went smoothly. We had only a minor gear problem, when the wire from our ham radio detached from our backstay, which we used for an antenna. Dan was able to quickly reattach it, and we were back on the air. On our Late-Leavers Net Jeff on *Mirage* reported that the lines in their spinnaker sock became tangled, and he had to hoist Cathy up the mast to straighten them out so they could take the spinnaker down—a scary job on a swaying mast in the middle of the ocean. A few days later their spinnaker wrapped around the forestay, and they had the difficult job, while standing on the bow, of unwrapping the spinnaker one roll at a time. Chuck on *Grey Eagle* reported that he tore several hanks out of his mizzen sail, and on *Furiant* the skipper reported his jib was torn in two during a squall. The boom vang fitting pulled out of the deck on *Oso Bueno* during a squall, causing an accidental jibe. Three boats lost the use of their autopilots. Only one, *Oso Bueno*, had a replacement, but they hated to use it because it was a power-guzzler, running down their batteries. Cruisers whose autopilots failed had to hand steer for hours, which can be very tiring. Bud on *Unity* reported that his refrigeration had failed, and on *Tres Locos* the Kaboda generator, which ran the refrigerator, failed. This meant eating the food immediately or throwing out all the refrigerated food they had planned for the passage.

People on several boats had success fishing by trolling a line behind the boat. We caught two tuna and a mahi-mahi. Bob on *Lemolo* caught a shark on his trolling line. Shortly afterwards he caught a mackerel, but it was eaten by a shark before he could pull it in. On *Kindred Spirits* they hooked two large fish—each about 250 pounds—on their trolling line. One of the fish dove deep and turned the boat around 360 degrees. Both fish were released. Jeff on *Mirage* reported that they wrapped their fishing line around the propeller, and Jeff had to dive under the boat to unwind it. Bob on *Lemolo* reported he had crossed the equator and had gone swimming. Just after getting back on board they saw an eight-foot shark following the boat.

Several people reported minor injuries. Bud on *Unity* fell down on the foredeck, hitting his head on the shrouds. Another cruiser injured his heel on a fitting on the foredeck. On *Mirage*, Jeff's back went out. Dave on *Oso Bueno* was bitten by a blue-footed booby twice when trying to force it to leave from its perch on their bimini.

The most serious problem was engine failure. Chris on *Infini* reported the first week out that he found a milky substance in his engine oil—probably water. A few days later John on *Skywater* reported his engine had failed. For five or six days in the doldrums both boats had no wind, and were drifting, afraid they would be swept past the Marquesas by the strong west-setting current. On the afternoon net Chris and John had a one-hour problem-solving session with Earl, a diesel mechanic, on *Spirit Wind*. Earl promised to meet them in Nuka Hiva to work on their engines. We could empathize with being stranded without the use of an engine. During our passage home from Hawaii we were becalmed for almost a week, with not enough fuel to motor through the extensive North Pacific high. Drifting around with no wind for days at a time can be terribly frustrating. By the third week another boat was engineless—*Tres Locos*. Chuck on *Grey Eagle* had a bad impeller, which caused his engine to overheat and burn out the muffler. The alarms went off as smoke and saltwater entered the bilge. He was able to repair the muffler with plywood and wire. *Tres Locos* reported he had fixed his Kaboda generator, and later got his engine going. Fortunately, the two boats without engines eventually were able to make it into port under sail. Cruisers were able to solve some of their problems underway, but would have to take care of more serious problems once they were in port.

Sharing their many problems on the net was therapeutic for cruisers. Although separated by hundreds of miles of ocean, cruisers didn't feel so alone, and received sympathy and helpful advice from others who had had similar experiences. After hearing about all the things that can go wrong on a passage we realized that we were incredibly lucky. We had no generator or refrigeration to fail, our autopilot and engine were both working, and our sails and rigging were intact. Our passage had been delightful and almost trouble-free.

## Deep-Sea Fishing

During our stay in Atuona we had checked into French Polynesia, bought lots of local fruits and vegetables, and shared many happy hours and landfall parties with friends who had arrived after us. We really enjoyed hearing about our friends' passages from Puerta Vallarta to the Marquesas. For most people, it had been their first passage, and they arrived with fabulous smiles. Although there were only 10 boats in the anchorage when we arrived, it was now getting crowded with new boats arriving daily, until there were 25 boats at anchor, and little room to swing. After 12 days at Atuona, we decided to leave for Hana Moe Noa, a small cove on the nearby island of Tahuata.

We raised anchor at Atuona and had an afternoon sail to Hana Moe Noa in light winds and warm sunny weather. Hana Moe Noa is a beautiful cove with a white sand beach at its head, and crystal clear, turquoise water. This was the favorite anchorage of Susan and Eric Hiscock who circumnavigated twice in *Wanderer III*. There is no one living ashore near the anchorage, just thousands of coconut palms. There were only six or seven boats at anchor, and it was very quiet and peaceful. The water was fairly warm, and it had some of the best snorkeling we had seen anywhere. We joined the crews of several other boats, mornings and afternoons, swimming around coral heads and rocks to look at the many colorful tropical fish, some the same as we had seen in Hawaii and Mexico, and some we had never seen before.

On our second day there Bobby and Gail invited us and Chuck and Marcia over to happy hour on their boat to discuss tides in French Polynesia. We had a hilarious time discussing everything but tides, but we did find out that Chuck had a good tide prediction program on his computer, and he loaned us a copy to put on our computer. That evening Dan tuned into the cruising net on our ham radio, and was listening to cruisers who were making passages when we heard a loud thump from our dinghy alongside our boat. We couldn't figure out what caused such a loud noise. Dan got out our powerful flashlight to investigate. We were amazed to see a two-foot barracuda thrashing around in our dinghy. It had probably leapt out of the water in pursuit of smaller fish, and accidentally landed in our dinghy. This was the easiest fishing we had ever experienced. It could have been dinner, but we were afraid of ciguatera poisoning, caused by a toxin in fish that live near coral. The toxin accumulates in fish that eat smaller fish, and since the barracuda is at the top of the food chain, we thought it best to return it to the sea rather than risk getting ciguatera. Dan grabbed it by the tail to avoid its sharp teeth, and tossed it overboard. We had a vegetarian dinner that evening.

The next morning, when we returned the tide program to Chuck, he talked us into staying another day so we could go deep-sea fishing on his boat along with Bobby and Gail and their son Cash. Chuck and Marcia had a lot of experience fishing in Florida, the Caribbean, and the Bahamas, and they assured us we could catch a mahi-mahi, a tuna, or a wahoo. We hoped to catch something big enough to feed the seven of us for dinner.

At 5:30 the next morning we were awakened by the sound of Chuck starting his engine. We quickly got dressed, grabbed some bread and bananas, and headed over to their boat, along with Bobby, Gail and Cash. We motored out of the small cove into the deep water and began trolling for fish. Chuck and Marcia had four poles, each with big reels of 30-pound test line, and dozens of lures to choose from. Chuck set up two poles with lures on the stern of his boat, and began slowly motoring, with the lines stretched out 200 feet behind the boat. We sat around in the cockpit talking, eating breakfast, and reading, while waiting for a fish to strike. Two hours later, as boredom was setting in, we had a strike. It had been agreed that I would get to pull in the first fish, so I took the rod and began slowly reeling in the fish, with a lot of coaching from Marcia, while Chuck ran the boat. "Let it run! Now reel it in slowly. Let it tire itself out! Keep the tip of the pole up." Dan helped me hold the rod up, and I reeled and reeled. The fish kept running the line out almost to the end, and I kept reeling it back in. This happened six times, and I was never able to get the fish even halfway up to the boat. After 45 minutes of this I was exhausted, my arms aching. Then this huge fish leaped into the air at the end of the line, and we all were shocked to see what we had hooked—a seven- or eight-foot marlin, probably weighing 200 pounds or more. It had an unmistakable torpedo shape and a long sharp bill for its upper jaw. It resembled the many marlin we had seen unloaded from charter fishing boats in Cabo San Lucas, except this fish was very much alive. I gave up. Cash, Bobby and Gail's strong 24-year-old son, gladly agreed to take over and try to bring the fish in. Cash was determined, and kept reeling the fish in for the next two and a half hours, during which time the fish ran the line out 30 more times while we watched in amazement. The fish made a couple more leaps out of the water—a spectacular sight. Cash still had the energy to keep going, while Chuck and Marcia cheered him on. Just when he had finally reeled the fish in fairly close to the boat, we saw a hammerhead shark cruising in the water behind the boat. Suddenly there was a strong pull on the line, we saw a huge splash at the end of the line and then it went slack. The line had broken at the knot at the lure. We didn't know whether the shark had attacked the marlin, or the marlin had made a spurt of speed, broken the line, and escaped. We hoped it was the latter. It was a great relief that it was gone. Dan and I had worried about bringing this huge marlin onto the sailboat where it could do extensive damage. We felt really sorry that we had tortured it for over three hours, dangling it behind the boat in shark-infested waters. When we discussed our adventure at the end of the day, Bobby, Gail, Cash and we decided it would have been better to cut the line when we saw what a huge fish we had hooked. Chuck and Marcia strongly disagreed. They were sure we could have brought it in if the shark hadn't intervened. It had been an exciting day, but it was our last deep-sea fishing trip. We didn't ever want to hook such a monster fish again.

## Tropical Fruit

After a three-week passage from Mexico to the Marquesas we had a strong craving for fresh fruits and vegetables. The fresh food we had purchased in Puerto Vallarta before departing was nearly all gone. In the tropical heat and with no refrigeration on the boat most of the produce that we hadn't eaten the first week or so had rotted and been thrown overboard. Only the hardiest vegetables had survived the trip—onions, potatoes, cabbage and jicama.

Our meals during the passage had been fairly tasty, consisting of canned tuna, chicken, beef, beans or lentils, supplemented with fresh vegetables and fruit. For the last couple of weeks we were eating mostly canned vegetables and canned or dried fruit. We were a little tired of cabbage and potatoes, however, and wanted more variety. We were also looking forward to some papayas, bananas, mangoes and especially pamplemousse, which we had fallen in love with on our first trip to the Marquesas six years previously. Pamplemousse, the French word for grapefruit, bears little resemblance to the grapefruit we eat in North America. It is green-skinned, about twice the size of normal grapefruit, with pale green sections, and a delicious sweet flavor.

When we rowed ashore at Atuona, we were fortunate to meet a young couple who picked pamplemousse and oranges from the trees in their yard to sell to the newly arrived cruisers. They sold us large quantities at very reasonable prices. At the small store in town we purchased a few vegetables such as tomatoes, lettuce and cucumbers, plus butter, eggs and baguettes. The locals had an abundance of delicious fruit. Whenever we ran out we were able to get pamplemousse, oranges, and lemons and even watermelon from the local people.

On May $2^{nd}$ we departed for the island of Fatu Hiva, a place we had wanted to visit on our first trip to the Marquesas. Because it is 45 miles southeast of Hiva Oa it is very difficult to sail there in daylight against the prevailing southeast trade winds. Amazingly, this time we had a couple of days of northeast winds, and were able to have a nice downwind sail. Leaving early in the morning and motoring for two hours until we were clear of Hiva Oa, we had a pleasant sail in 10-to-12-knot northeasterly winds and a favorable current, arriving in port well before sunset. The bay on Fatu Hiva is a spectacular anchorage, with tall spires of black lava rock lining the head of the bay and lush green hills all around. Once again we were running out of fresh food, so we walked into the village to see what we could find. The villagers on Fatu Hiva preferred to trade for their fruits rather than sell them. We traded balloons and chocolate bars with the children for bananas and limes. Friends of ours got a real bargain when they traded a Dremel tool set and an extra skillet for a large tapa cloth, and a carved paddle. We traded a homemade string bag, string, netting needles and instructions for making a string bag for a large fish. A young woman we met had papayas, but she wanted clothing in exchange. The next day we brought her a purple sleeveless blouse in trade for a dozen papayas. When we attended church service we were surprised to see her in the congregation, looking beautiful in her white slacks and purple blouse against her tan skin.

Thirty days after entering the Marquesas we sailed to Taiohae Bay on the island of Nuka Hiva, which has the largest village in the Marquesas. We were able to purchase lots of vegetables at the store, and also beer (at $4 a bottle we didn't buy much).

The last island we visited was Ua Pou, our favorite Marquesan island on our trip six years before. We bought lots of produce in the store, and even had dinners ashore at small restaurants twice during our stay. Jeff and Cathy from *Mirage* joined us for our treks into town. We met a retired French submariner who had a vegetable garden similar to what he might have had in France. He sold us tomatoes, cabbage, Chinese cabbage, green beans, and breadfruit. A young man, Peter, whom we met on a walk into town, had lots of fruit trees in his yard. He picked lemons, bananas, papayas and pamplemousse for us in exchange for a large fish lure and some fishhooks.

A few days later, when we and Jeff and Cathy were getting ready to depart for the Tuamotus, Jeff and Cathy and I walked into town in search of mangoes. We wanted fruit for the four-day passage and also for our stay in the Tuamotus, where the only fresh food available is coconuts. Dan stayed on the boat recovering from a cold. During our walks around the village we had noticed huge trees in people's yards loaded with mangoes. At several houses we knocked on the door to ask if we could buy some mangoes from their tree. Unfortunately, no one was home. Then we spotted a huge tree loaded with mangoes in the backyard of a Catholic church. We got up the nerve to knock on the door of the small parish house behind the church. Cathy, in her halting high school French and some pantomime, asked the French priest if we could buy some mangoes for our trip to the Tuamotus. The priest listened to her request, but told her (in French) that the man responsible wasn't there, but we should come back in 10 minutes. When we returned to the church the priest gave us permission to pick a dozen mangoes. They were high up in the tree, but Jeff was able to knock them off with a long pole, and Cathy and I caught them as they fell. They would be a great treat on our upcoming passage to the Tuamotus.

The next morning, when we arrived at the church to attend the Sunday service, we watched a long parade of young boys and girls dressed in white led by a bishop in splendid red and white robes, carrying a large gold scepter. We learned from bystanders that the young people were being confirmed, and the bishop from Tahiti was conducting the ceremony. The bishop bore a strong resemblance to the priest from whom we had requested the mangoes the day before. It dawned on us that this was the priest we had talked to at the parish house. As a visiting bishop he couldn't allow us to take mangoes from the tree without asking the local priest. He had probably never had a request like ours. We were grateful he had understood our request and received permission from the local priest for us to pick them.

## Landfall in the Tuamotus

We left the anchorage on the southwest side of Ua Pou on June $9^{th}$, headed for Raroia in the Tuamotus, a distance of 420 miles. The sailing was pleasant the first day with a 15-to-20-knot wind behind us. We made 114 miles the first 24 hours. The second day we hit a series of squalls with heavy rain and gusts of 30 knots or more. Dan had to reef the main and roll in the jib during the squalls, then roll out the jib again when the wind decreased. The seas became very rough making it difficult to sleep and almost impossible to cook. The weather was much the same on the third day, but we continued to make good progress. By the fourth day the weather improved and we made landfall at 8:20 am. By luck, the timing was perfect for going through the pass—two hours after the low slack at Rangiroa. There was an ebb current of two to three knots, but we were able to motor through the pass easily.

Once inside the lagoon the seas were flat and we could motor to the other side where five boats were anchored near a large motu. We were very happy to be in a calm, peaceful anchorage where the only land in sight are low-lying motus covered in coconut palms. It is beautiful, but remote. Everyone has to be self-sufficient. We had solar panels to keep our batteries charged up, a watermaker to convert sea water to fresh water which we could run on sunny days when it would not deplete our batteries, and almost six months' worth of canned, packaged and bulk food that we had purchased in Mexico. Everyone stocks up with more food than they think they can use because the cost of food in French Polynesia is prohibitive. In the cool of the mornings I usually baked cakes, cookies, muffins, home-made granola or bread, or some combination.

We knew most of the cruisers anchored near by. As soon as we anchored our friends Jeff and Cathy on *Mirage* invited us over for fish chowder from a grouper they had just caught in the lagoon, and later everyone in the anchorage was invited ashore to share a chocolate cake to celebrate Tam and Sallee's third wedding anniversary. We had a great time visiting old friends and new ones.

One morning we went ashore with friends to walk across the motu at low tide to look at the tide pools and see the great expanse of ocean outside the lagoon. That afternoon Sallee, on *Temptation*, showed a group of us how to weave baskets out of coconut fronds, crisscrossing the long narrow leaves on each side of the stem to form a strong basket, such as the locals use for carrying fruit. Most afternoons we took our dinghies over to nearby coral heads in the lagoon to snorkel and spear fish, using spears we had purchased in Mexico. Dan and I had never tried spear fishing, but our friends claimed it was easy. We were hoping to catch a grouper for dinner. We dove down about six feet and Dan took aim at a moderate size grouper that was in plain view at the base of the coral head. As soon as he pulled the spear back, the rubber tubing that was attached to the end of the spear broke. Now I decided to try with my spear. I looked at the fish, but couldn't bear to stab it between the eyes. It seemed heartless to kill this sedentary fish that was staring at me completely unafraid. Fortunately, we were with Jeff and Cathy who were experts at spearing fish, and they gladly speared an extra one for us. From then on they always stopped by to ask if we wanted fish before they left on their fishing expeditions, and brought back one for us if we wanted it. We enjoyed snorkeling on the coral heads to look at and identify the many colorful reef fish, but never tried spear fishing again.

Twice after successful fishing trips all the cruisers gathered in late afternoon on the motu for a potluck dinner. We made a campfire of coconut fronds gathered from the beach and had a variety of seafood—grilled grouper, curried grouper, grouper kabobs with peanut sauce, octopus, parrotfish, and scallops, supplemented with rice and bean dishes, and cookies, cakes, or brownies.

After days of sunny, balmy weather, which was pleasant due to the constant trade wind breeze, we decided to leave for Kauehi, an atoll about 150 miles to the north. Ten days after entering Raroia we raised anchor and motored across the lagoon towards the pass. Suddenly the sky turned dark and we could see squalls to the west— not good conditions for maneuvering through a narrow pass and sailing to another atoll. We turned around and headed back to the anchorage to re-anchor and wait for better weather. It blew hard from the west and rained all night. The next morning the rain stopped, but it was still overcast and gusty. We visited with a Swedish boat, *Rollon*, anchored near by, with Lars, Karen and their 18–month-old daughter, Maria, on board. They were also waiting for favorable weather to leave Raroia. We traded weather forecasts we had gathered from different sources, and talked about ham radio. Lars was a ham who wanted to be able to communicate by Morse code, and we had an extra Morse code key on board that we sold to him for a bargain price.

The next morning the skies cleared, and we and the crew of *Rollon* raised anchor to motor out of the pass. *Rollon* preceded us, and once outside the pass Karen called us on VHF radio to ask if we wanted a loaf of bread. She had baked several loaves that morning and it was fresh out of the oven. What a delightful surprise! They slowed down so we could pull alongside their boat, Dan stood out on the side deck as they passed us the loaf in a plastic bag at the end of their boat pole. It was delicious Swedish rye bread which we consumed immediately while it was still warm as we began our sail to Kauehi.

Our visit to Raroia was unusual in that we didn't see any local inhabitants. Without the company of other cruisers we would have felt isolated and lonely. Sharing the delights of this atoll with a congenial group of cruisers made our stay on Raroia memorable.

## Kauehi

Some of our friends were anchored at Kauehi, and on the radio they gave glowing reports of the peace and beauty of this atoll. Our trip to Kauehi was only 160 miles, an easy overnight sail. We left Raroia in the morning and arrived at the pass mid-afternoon the next day. We had missed the low slack tide in the morning, but decided to go through the pass during the high slack tide in the afternoon. There were some ripples in the pass, but it looked fairly calm. The current was only two knots in the direction we were headed, so we had no problem.

Once inside the pass we had to motor about two miles across the lagoon to get to the anchorage where three other boats were anchored. Our friends told us there were a few scattered coral heads in the lagoon, but they were easy to see. I went to the foredeck to look out for coral heads. With the sun at our backs, and with polarized sunglasses, it was easy to see through the clear water. Occasional rain squalls hit us as we motored across the lagoon. Then the sun disappeared behind a black cloud, and I could no longer see anything underwater, so I gave up and returned to the cockpit. This was a mistake, as shortly afterwards we felt a jolt, and heard a sickening "scrunch" as we hit something underwater while traveling at full speed. The boat seemed to rise an inch or so, but then continued on. Looking over the side of the boat we could see the distinct shape of a six-foot wide coral head, resembling a huge yellow cauliflower, fairly near the surface of the water on our starboard side. We had grazed the left hand edge of the coral, and probably knocked off a small piece of it. We suspected it had scratched a long gouge near the bottom of our hull. It was frightening how close we had come to going aground on this large coral head, which would have happened if we our track had been a few feet further to the right. We continued on for 10 more minutes to the anchorage, relieved we had arrived intact, and dropped our anchor in 36 feet of water, near the other anchored boats. We went to bed early, tired after our overnight passage with little sleep, but were awakened in the middle of the night by rain squalls and 30 knot winds. We were happy to be safe at anchor in the peaceful lagoon instead of sailing on the ocean.

The next day Dan dove under our boat to assess the damage from hitting the coral head. He found a small chunk of fiberglass had been knocked out from the bottom of the hull. Fortunately the damage was minor. The fiberglass in this part of the hull, containing the encapsulated lead ballast, is probably an inch thick, and the hole left from hitting the coral was less than a half inch deep. As a precaution Dan prepared a handful of underwater epoxy and filled in the hole. We could repair the damage permanently when we hauled the boat out it New Zealand.

That afternoon Chuck, from *Gray Eagle*, came by with some strips of grouper skins and suggested we put them on the hooks of our wiggle-worm lures and try fishing. Dan tried fishing from the bow of our boat and caught a nice size grouper on his first cast. I tried later and caught a good size oio (bonefish). Dan filleted my fish, and we gave one of the oio fillets to Heiko on *Carpe Diem*, and pickled the other fillet. We had grouper teriyaki for dinner that evening.

The second day Chuck and Marcia invited us all over to *Gray Eagle* for a spear fishing expedition. They had taken us fishing at Hana Moe Noa in the Marquesas, where we hooked a huge marlin. We brought our snorkeling gear and a lunch and rowed over to their boat in early morning. Jeff and Cathy from *Mirage*, and Heiko and Kirsten from *Carpe Diem* joined us. Chuck motored out and anchored near a large coral head in the lagoon where we tried snorkeling, but didn't see any grouper. Chuck raised anchor and motored further out to another coral head near the pass. Dan trailed his lure and caught a good size snapper along the way. After we were anchored in the lee of the large coral head we put on our snorkeling gear and investigated the sea life. The coral head was beautiful, with spiral tubeworms and a large variety of small reef fish. There were also many sharks circling the base of the coral reef—grey reef sharks and black-tipped sharks. My instinct was to get out of the water as quickly as possible, but the other cruisers, who were experienced snorkelers, told us sharks wouldn't attack people at a place where there were plenty of reef fish for them to eat. Nevertheless, I was extremely nervous, and kept an eye on the sharks the whole time as they swam back and forth. Then Heiko speared a large grouper. Both he and Kirsten swam like mad back to their inflatable dinghy, trailing the fish behind them on his spear. We all climbed out of the water as quickly as possible, knowing sharks would be heading towards us, attracted by the bleeding fish. When Chuck motored back to the anchorage we caught a second snapper. Marcia cooked us all a delicious cioppino of fish, scallops and canned baby shrimp for dinner. We stayed on board for a few hours hearing thunder and watching a spectacular lightning display. It blew 20 to 25 knots after we returned to our boats and we saw lightning till midnight.

The third day we all went to *Gray Eagle* in the morning to have coffee, exchange recipes and addresses, and future plans. We had pickled the fish we caught two days before, and shared some with the other cruisers. Dan caught a nice size grouper from the dinghy in the afternoon, and we had Jeff and Cathy over for a dinner of grouper with yogurt sauce. The fourth day I tried fishing from the dinghy and caught a 15-inch grouper immediately. Jeff and Cathy invited us aboard their boat for a dinner of curried fish using our grouper.

Dan and I left the next morning, motored to the small village near the pass and anchored in front of the motu. It was July $2^{nd}$, and we wanted to get to Tahiti by Bastille Day to see the festivities. At the small store we were able to re-supply with eggs, cheese, butter, crackers and sugar. We met Ernest, the head of the post office there, who told us there were 55 to 65 people living in the village. The name of the village is Tearavero, which he explained means "keep fishing", a very apt name. We asked if any of the fish were toxic—he said only the large grouper (the same fish we had been eating for the last four days!) He gave us a tour of the Catholic church, which had beautiful embroidered tapestries hanging on either side of the cross. We gave him a small package of chocolate chip brownies I had baked and some balloons for the children. He told us we should check out the next morning before we left.

We went ashore about 9 am to find the police chief and have our passports stamped, but found out he had gone to another atoll, Fakarava. We then went to look for the chief of the village, but his wife told us he was visiting the pearl farm in the lagoon. It didn't matter, since checking out is only a formality, but we were disappointed that we didn't have a stamp in our passports as a record of our visit to Kauehi.

For five days we had enjoyed excellent fishing and sharing fish lunches and dinners with our fellow cruisers. Fish were plentiful in the lagoon, and were extremely easy to catch. However, if anyone had told me we would endure numerous squalls, thunder and lightning, hit a coral head in the lagoon, eat possibly toxic fish, and swim with reef sharks, I would never have visited Kauehi.

## Rat on Board

After our visits to Raroia and Kauehi, we made an overnight passage to Takaroa. Our reason for choosing to visit Takaroa was to see Ah Sahm and his family whom we had visited at their pearl farm six years earlier. After entering through the pass we tied up to the nearby wharf and went ashore to walk around the small village and check out the general store. We met the local meteorologist, who invited us to see his modern weather station of which he was quite proud. He told us that Ah Sahm and his family had left Takaroa and moved to the neighboring atoll of Apataki because of a land dispute. We were extremely disappointed that we would miss visiting with them. Since Takaroa otherwise held no interest for us, we decided not to anchor in the lagoon, but to stay at the wharf overnight. The next morning we would leave for Tahiti when we could go through the pass at slack tide.

That evening we had a simple dinner of canned chili and rice, and I baked a pan of our favorite dessert, chocolate chip brownies. Before a passage I usually baked a cake, cookies or brownies so we would have a homemade dessert to supplement our usual fare of mostly canned food. We were tired after our 24-hour passage from Kauehi, so Dan left the dishes in the sink, knowing he would have time to wash them in the morning while we waited for slack tide. It was very warm that evening, so we left all the ports open to get some breeze.

I was awakened when it was still dark, around 3:30 am, when I heard a metallic noise like the rattling of a pan in the galley area. I woke Dan, asking "What's making that rattling noise?" Dan replied "It's just the rocking of the boat causing the dishes to rattle in the sink." It was completely calm at the wharf, there were no ripples on the water, and I suspected it was something else. I searched around with a flashlight and discovered a hole gouged in the middle of the brownies in the pan I had left on the chart table. The hole had tooth marks on the edge. Oh horrors! We must have a rat on board. Rats like to hang around wharves to chew on the copra (dried coconut) which is stored in burlap bags waiting to be picked up by the next copra boat. Rats can gnaw through electrical cords, or through plastic tubing. If the tubing is connected to a through-hull fitting, saltwater can pour in and sink the boat. We had to catch the rat before it could do serious damage.

When Dan got up to wash the dishes early in the morning, a large, brown Norway rat leapt out of the dish cabinet and hid behind the stove. The stove is gimbaled, so I had the brilliant idea that we could kill the rat if we gave the base of the stove a quick push, crushing him between the stove and the stainless steel surrounds. That was a colossal failure—as soon as we tried it the rat dashed out from behind the stove, across the floor and up into the quarter berth. Our quarter berth is a seven-foot long by three-foot wide bunk next to the engine compartment. It was no longer used to sleep a third crewmember, but instead the space was crammed to the deck level with all of our spare equipment, fog horns, life jackets, safety harnesses, winter clothes, extra charts, a box of files, a sewing machine and my violin. The only way we could kill the rat was to entice him out of the quarter berth and catch him in a trap. We rigged up a makeshift trap using a cardboard wine box with a trap door, put it on the floor near the quarter berth with a brownie inside and waited quietly. We could hear the rat rustling around in the quarter berth, but he didn't venture out. By this time it was after sunrise, and we knew we'd have to wait till evening when it would be active again to try to catch it.

After breakfast we went ashore to the general store to buy a rat trap. Fortunately the shopkeeper had just one left for about $5. We were really grateful, knowing that this was our only hope of catching the rat. We had to give up our plan of leaving for Tahiti that morning. Starting a three-day passage with a rat on board, gnawing away at tubing or electrical wires during the night, would not be wise. Instead we left the wharf and motored out into the middle of the lagoon far from shore. There we could leave the ports open overnight without the fear of another rat boarding the boat. It was a beautiful, clear day, and the water was warm enough for swimming, so we spent an enjoyable afternoon snorkeling to observe the colorful tropical fish swimming amongst the small coral heads in the lagoon. After dinner Dan worked on the new rat trap, adjusting it so that the slightest motion would cause it to snap. I mashed one of the brownies into a small cube as bait for the trap, and we put the trap on the chart table, where we had left the pan of brownies the night before. We went to bed about 8 that night just after sunset. Before we had even fallen asleep we heard a loud snap—Hooray, we got him! Oh no, he was still alive! The trap had caught him on the hindquarters. Dan wanted to throw the rat and trap overboard, but I insisted that Dan kill him. I found a short piece of two-by-four amongst our spare scraps of wood in the forepeak. Dan gave the rat and the trap a couple of hard whacks, and when it stopped moving he released the dead rat into the lagoon. What a huge relief! We threw the rest of the brownies into the garbage.

## Protests in Polynesia

Our sail from Takaroa took three days, but we had pleasant conditions and arrived in Papeete, Tahiti a few days before Bastille Day. Before we left the Tuamotus, we heard from other cruisers on the radio that protestors had blocked all roads into Papeete. The French had to ferry workers into the city from outlying districts. The local Tahitians were protesting the resumption of nuclear testing on Mururoa. The French officials had prohibited Greenpeace's *Rainbow Warrior* from entering Papeete Harbor.

On Bastille Day we took Le Truck into the city to watch the big parade. The French officials had finally relented and allowed the *Rainbow Warrior* to come into harbor. There was much cheering by the locals as it arrived in Papeete. Instead of watching the usual French parade of French troops and tanks we found a large gathering of Tahitians, many wearing anti-bomb T-shirts and waving blue and white independence flags. Some carried banners denouncing France and President Jacques Chirac. They had set up large loud speakers near the harbor on which they played music and gave lots of speeches in Tahitian. The protesters paraded up and down the long wharf in the harbor where two huge French navy ships were moored. They carried posters denouncing the bomb testing. The crew on the navy ships ignored them.

After watching the protest on the wharf we walked over to the tourist bureau a few blocks away. This is on the main street where the military parade was scheduled to pass by. Greg, Joanne and Hanna from *Asia*, who had come for the Bastille Day celebrations, joined us. We tried to figure out what was happening. In a short while we saw a couple of Tahitians lie down on the pavement of the main street, joined soon after by a large crowd of men, women and children with their blue and white independence flags. The men and women sat down in the middle of the street, completely blocking the parade route through town, although they allowed normal traffic to pass by. It was a peaceful, but noisy demonstration, accompanied by the men blowing conch shells, and everyone beating sticks and rubbing half shells of coconut against the pavement. It reminded us of the anti-war protests we had been to in Berkeley and Eugene in the 60s, but much more colorful. The protest continued until noon, and still there was no military parade.

We learned later that the French military, not to be thwarted by the locals, staged a very short parade of tanks and military vehicles through some side streets, well away from the demonstrators. However, the news on French television showed videos of the parade through the streets, with no mention of any protests. Thus the French officials hid the fact from the world that there was a great deal of local opposition to their bomb testing in French Polynesia.

We, and most cruisers we know, were sympathetic to the cause, and were appalled that the French were determined to resume testing despite world opinion and strong local opposition. The independence movement, called Hiti Tau, started some years ago, but had gained strength due to Chirac's decision to resume nuclear testing.

We learned about the *Rainbow Warrior*'s protest at Mururoa through Australian radio broadcasts. When they approached the atoll four crew members went ashore on the beach in well-equipped inflatable dinghies. Three were arrested immediately, but one was thought to still be hiding out with enough food and supplies to last for weeks. When the ship tried to enter the Mururoa lagoon it was boarded, and the French broke into the cabin using tear gas. The crew were detained briefly, then were returned to the *Rainbow Warrior* by the French gendarmes. The French towed the *Rainbow Warrior* outside the 12-mile territorial limit, and closed the entrance to the lagoon by putting a chain across it.

Over the next few weeks we heard on Australian Southwest Radio that the *Rainbow Warrior* proceeded to the harbor at Rarotonga, in the Cook Islands, after leaving Tahiti. The Cook Islanders held the largest protest they had ever had against the French resuming testing. In spite of all the protests the French tested three nuclear bombs on Mururoa, spreading radiation over a large area of the nearby atolls. The French officials maintained the tests were harmless, but we wondered why, if the bombs were harmless, they didn't test them in their own backyard in France.

## Playing Music on Moorea

After our chance meeting with Lynn and Rob in the Sea of Cortez, we got together with them again on the island of Moorea in July. In the meantime we had met up with Greg, Joanne and her daughter Hanna on the sailboat *Asia*, from Seattle. Greg was a music teacher in Shoreline, and played a variety of instruments, including guitar, saxophone and the flute. He and I had joined together to play flute-violin duets in June in the Tuamotus. Then we had spent a couple of weeks in Tahiti during the Bastille Day festivities in mid-July. Towards the end of July a young couple, Laura and TJ, were planning to get married on the nearby island of Moorea. Greg had offered to provide music for the wedding. Greg had some trio music suitable for a wedding, so Lynn, Greg and I got together to practice several short pieces and the wedding march for two flutes and a violin. It was to be an outdoor wedding at the anchorage of Opunohu. Our friend Bobby, on the boat *Tulum III*, officiated—he had obtained a license to perform marriages from a new-age church in California. When we arrived at Opunohu there were already 18 boats at anchor. *Scout* was beside us and *Asia* came in and anchored behind us. The cruisers left their dinghies on the beach and walked up to a large coconut grove. It was sunny, but pleasantly cool with a gentle trade wind breeze. Greg, Lynn, and I set up our music stands on the edge of the grove and played a couple of pieces while a crowd of about 40 cruisers arrived on the beach, and gathered under the coconut palms. We began playing the wedding march as Laura and TJ walked up from the beach. The bride was dressed in a beautiful pareau, the groom in shorts, a colorful shirt, and both wore sandals, frangipani leis and flowers in their headbands. Bobby read the vows he and Gail had composed, in lieu of the traditional vows. After the ceremony we played a couple more pieces. Lynn was quite nervous, having never performed before, but it went well. Everyone gathered on the beach after the wedding for a wonderful potluck.

I had played in a string quartet at several wedding receptions in Seattle in previous years, but never at an outdoor event in such beautiful surroundings. Although none of us knew the couple everyone enjoyed participating in this happy celebration.

## Hiking to the Top of Rarotonga

After two and a half months visiting 13 islands and atolls in French Polynesia it was time to depart for Rarotonga, located about 500 miles southwest of Bora Bora. Rarotonga is the largest of the Cook Islands, although less than half the size of Bainbridge Island. When we checked out of Bora Bora we were given back our bond money of about $3,000, and we decided to have a going away lunch at the Bora Bora Yacht Club with the crew of three other boats who were also about to depart. The hamburgers were good, though we thought they were a rip-off at $11 each, and the waitress was hopeless, mixing up the hamburger, cheeseburger and fish burger orders so no one got what they ordered, and forgetting to bring the Sprite. She put two Sprites on the bill, and then agreed to take one off. She said she couldn't deal with giving us four separate bills, so we had to sort it all out for ourselves amongst much hilarity.

The next day we stowed everything, and Dan and I had a refreshing swim from *Shaula* in the crystal clear water before departing. It was not an easy passage. We were plagued with light winds or no winds for the first two days, making it necessary to motor for long stretches. On the first day I was sleepier during my three-hour watches than on my off watches, and couldn't stay awake during the 10-minute intervals between looking around for ships. I had frequent nightmares about a bright light of a ship heading directly towards us, or of our boat slamming directly into a high island. On the third day we had too much wind in the wrong direction. The seas were rough, occasionally breaking over the bow of the boat, and then we ran into a series of squalls that lasted all afternoon and evening. On the fourth day the winds moderated and we had a nice downwind sail for most of the day. Then the wind died, and we had to motor again. The next day we were sailing in strong gusty winds, with dark clouds all around. We were overjoyed to see the tall island of Rarotonga rising above the horizon in the early afternoon of the fifth day.

We pulled into the long wharf along one side of the harbor and tied up with our yellow quarantine flag hoisted below the spreaders. Soon the health officer came along, boarded our boat, had a beer, then sprayed the inside of the boat with insecticide to kill any flies, ants, cockroaches or other insects. The next morning we went to the harbormaster's office to fill out the forms for customs and immigration. The wharf is reserved for large ships and for incoming sailboats, but once cleared into the country sailboats had to move to the very crowded anchorage. After the harbormaster assigned us a spot we spent the rest of the afternoon maneuvering into our assigned space amongst the 25 other boats at anchor. The procedure is to drop an anchor well in front of the assigned buoy, then tie stern lines to two buoys so that your boat doesn't swing around. We were tightly packed into the small anchorage, with anchor chains crisscrossing each other and stern lines going everywhere, resembling a giant underwater spider web. One afternoon we heard a loud thump against our hull and found that a nearby boat had dragged anchor and banged into our boat. Dan helped the skipper set a second anchor to hold the boat in its assigned space. When a boat near us wanted to leave we had to stay aboard and help untangle our anchor chains. It was a calm anchorage in the prevailing southeast trade winds, but we could imagine the chaos if the wind switched to the north and everyone had to leave at once. Fortunately that didn't happen during the two weeks of our stay.

After 18 months in Mexico and several months in French Polynesia we were delighted to be in an English-speaking country again. The Cook Islands are a New Zealand protectorate, and this island, though small, had many restaurants, pizza stands, a museum, a library, and a movie theater. Buses circled the island all day both clockwise and counterclockwise on a paved road. Tickets were $3, and you could get off whenever you wanted and re-board the next bus that came along. We hadn't seen a movie for two years, and we made up for it by seeing three movies in our short stay for only $2 a ticket. We had pizza, ice cream, and good restaurant meals, all at reasonable prices. It was a little piece of paradise. When we weren't out enjoying what the island had to offer we went to happy hour on friends boats, and twice had a potluck on shore near the harbor with 40 or 50 cruisers. The only disadvantages of being on such a civilized island was that we had to endure the loud music coming from the open air bar near the harbor every night till 1 am, and the roar of Boeing 747s that took off and landed daily at the international airport near the harbor.

One afternoon we and friends from three other boats decided to hike across the island. The hike climbs to the summit of a 2,000-foot mountain in the center of the island. The very steep trail is crisscrossed with a maze of exposed tree roots. It was a strenuous climb, in which we had to step on and grab roots all along the way to pull ourselves up. In all my years of hiking in the Smoky Mountains and the Sierras I had never hiked a trail like this. At the top is a narrow vertical rock called the needle, which none of us climbed, but at the base of the needle we had spectacular views of the ocean and surrounding peaks. Hiking down the other side of the mountain was challenging also. We had to climb down two ladders on the steepest parts of the trail. When we reached the bottom we were rewarded with a nice level trail along a stream to a beautiful waterfall. We were all exhausted by then, and still had the prospect of a long walk along the road halfway around the island back to the harbor. Fortunately, we spotted a telephone truck parked on the road and one of our friends asked the driver if he could give us a ride. Ten of us piled into the back of the pickup truck and two in the cab, and the Telecom employee drove us all the way back to the harbor. It was probably strictly illegal for him to take passengers, but we promised not to sue in case of an accident. We got back in late afternoon and had a delicious meal at a pizza stand, followed by ice cream. I had a free hot shower near the harbormaster's office. The water was heated in a boiler above a wood fireplace, the scraps of wood furnished by a local carver, and all that was asked was that we added wood when we finished to keep the water hot for the next person to use.

Another day we and Gail and Bobby from *Tulum III* took a bus halfway around the island to visit Arnold, the ham radio operator who gave daily weather forecasts for cruisers in the South Pacific. He told us he had been a pilot in the New Zealand Air Force and had nearly met his maker when the weatherman didn't advise him of a low-level weather pattern. He and another pilot were caught in downdrafts and nearly lost their planes. Arnold vowed he'd prepare his own weather reports, and learned to put together reports from Fiji and New Zealand to make very accurate weather forecasts for the

South Pacific. He was sent to Rarotonga as a communications technician, married a local, and had lived there for more than 30 years. He was an excellent weather forecaster, and we relied on him for his daily forecasts of winds and seas on the ham radio net, telling us when it was safe to make passages and avoid storms. We all greatly admired him for his contribution to our safety at sea.

Several ships came into the harbor the first week we were there and tied up to the long wharf. They had been to Mururoa in French Polynesia to protest the French nuclear bomb tests. One was a large Fijian ship with nuclear protesters from nine countries, and had arrived after weeks of drifting at sea with a broken-down engine. We saw the ship being towed into the harbor by a local patrol boat so they could tie up to the wharf. We could imagine the relief of the South Pacific islanders at finally reaching port. Most of the passengers had probably never been to sea before. We watched them walk down the gang plank to shore and kneel down to kiss the ground. I could empathize with them. I've felt this way after every rough passage we've made, being very eager to set foot on solid ground, although I've never actually kissed the ground.

When we visited Papeete, Tahiti, two months before we had observed the movement for independence from France and the protest against the French resuming underground bomb tests at Mururoa, but the protests were peaceful. Since then the French had conducted an underground bomb test at Mururoa, and the protesters became aggressive. We learned from a cruiser in Papeete that the airport there had been burned to the ground, and numerous fires set in the industrial area. Thirteen people were injured, six of them gendarmes. This was a very stressful time for the Pacific Islanders who wanted to protect their island from nuclear testing and whose views were not being listened to by the French authorities. They had our sympathies, although we didn't approve of all their methods of protesting.

In the next few days a 70-foot Fijian ship and a huge New Zealand survey ship came into the dock. The New Zealand boat was awaiting the arrival of a cargo plane with supplies they were taking to the protest fleet in Mururoa. Rarotonga seemed to be the departure and re-supply port for the protest boats.

Towards the end of out stay at Rarotonga we heard a young lady single-hander call on the VHF radio requesting a tow into the harbor. Two dinghies from anchored sailboats went out immediately, but their outboard engines didn't have enough power to tow the sailboat. The single-hander, Linda, who was exhausted, made the mistake of dropping her sails before help arrived, and was helplessly drifting. Finally, two friends of ours went out as crew in a charter fishing boat that had an engine large enough to tow the sailboat. They tied their stern to the bow of the sailboat with a half inch diameter towline, which broke shortly after. Then the guys in the two inflatable dinghies helped tie a heavier line to the sailboat. The sailboat was nearly lost when the second towline slipped off the cleat and the sailboat drifted towards a reef near the entrance to the harbor. Finally, the line was re-connected and the boat was towed off just before hitting the reef. When the sailboat arrived at the wharf Dan and about 15 other cruisers went over to help with the dock lines. Linda told her rescuers that her self-steering had failed, so she had been hand steering in very rough seas for two days, and her engine had failed twice. She said little else before she went down below to rest and recover from her ordeal, perhaps reconsidering her decision to sail around the world single-handed.

A few days later we began provisioning for our next passage—either to Niue or Tonga. The weather was still too windy and rainy to leave, but we were hoping for a weather window in the next couple of days. We shopped in the morning for fresh food at the farmer's market and at the Cook Island Trading Co. store for bread, potatoes, eggs, lentils, etc.

Rarotonga, this very small island in the middle of the South Pacific, had provided us with much enjoyment and many unforgettable experiences. We were reluctant to leave, but had other islands to visit before the cyclone season began in a couple of months.

## Niue

We left Rarotonga in the Cook Islands hoping to stop at the small island of Niue on the way to Tonga, but only if the winds cooperated. It would be a small deviation from the straight line route between Rarotonga and Tonga. We hadn't committed ourselves to stop there, so I hadn't bothered to sew a courtesy flag, as I had for French Polynesia and Rarotonga. It was the last week of September, so we still had two months to reach New Zealand and avoid the tropical cyclone season, which begins in December.

Niue is a small raised coral island with a population of only 2,200 people, and very few tourists. Most of the tourists come from New Zealand, which supports the country. The residents of Niue have dual citizenship with New Zealand. Most of the young people leave to go to New Zealand after high school because there is little to do on Niue except grow vegetables or raise cattle. The population of Niueans in New Zealand is about 10 times the population on Niue. Unlike Hawaii with its five islands and French Polynesia with its myriad of islands and atolls, Niue is just a single circular island hundreds of miles from any other land.

During our five-day passage we had moderate winds and easy sailing interspersed with a half day of motoring when the winds died. We spotted the island on the horizon at dawn, and proceeded to the anchorage, which is an open roadstead. The anchorage is safe in the prevailing southeast trade winds, but would become dangerous if the winds switched around to westerlies. Westerly winds are rare, and fortunately didn't occur while we were there.

There were very few ships in the seas between Rarotonga and Niue, but we did see another sailboat heading west as we approached the anchorage. We talked on the radio to the cruisers who told us they were departing for Tonga. They loved Niue, and had stayed for eight days. It was our good fortune to be able to tie up to the mooring buoy they had just left—all the others were taken. Anchorage is difficult there because of the great depth (70 feet), but there are six mooring buoys along the shore available on a first-come, first-served basis.

We were amazed by the clarity of the water—we could see the bottom 70 feet below us. The waters were teeming with poisonous sea snakes, however—black and white banded slender snakes swimming gracefully near the surface. They are supposed to be harmless, having such a small mouth that they can't bite people, but I still couldn't bring myself to dive into the water to swim, imagining that their mouths might be large enough to bite one's little finger or toe. The idea of swimming with poisonous snakes makes my skin crawl. Dan doesn't believe in imagining the worst case scenario. He pooh-poohed my fears and dove in shortly after we arrived and had a refreshing swim in the warm, clear water.

Niue is the only island we had been to where the dinghy landing is 15 feet or more above sea level, making it a chore to get onto land. You row your dinghy up to the huge concrete wharf, where one person has to climb up a tall vertical ladder on the side of the wharf. That person swings a hook from a crane mounted on the edge of the wharf over the water and lowers it down to the dinghy. The other person attaches the hook to a bridle on the dinghy, and climbs up the ladder to the wharf. Then you crank the dinghy up to above the level of the wharf, and swing it around to lower it onto the concrete wharf, and detach the hook for the next person to use. The reverse process is used when you leave to go back to the boat. We soon tired of this procedure, so once ashore we usually stayed all or most of the day to avoid having to repeat the operation.

When we first went ashore we learned that we could rent mountain bicycles all day for only $5. We asked for a lock, but were told that we didn't need one—there had never been a theft on the island. You can leave the bikes anywhere along the road without locking them. We rode our rental mountain bikes part way along the 36-mile road that circles the island, and every few miles we left the bikes on the side of the road while we walked down the narrow paths to the sea to see the fantastic coral arches, deep chasms, saltwater pools ideal for snorkeling, and beautiful white sand beaches, all for free. It took two days to see all of the sites that were easily accessible.

On one of our forays ashore we learned that the trimaran that was tied to a buoy near our boat had come loose and begun to float off. Fortunately, some cruisers saw it starting to disappear and were able to chase it with their dinghies and tow it back to tie it temporarily to one of the sailboats on another mooring. We had seen the owners that morning traveling around the island on their rental mopeds, completely oblivious to the fate of their boat. The owner had tied his trimaran to the mooring with polypropylene rope, which is too slippery to hold a knot. They had to return that afternoon and re-tie to the mooring with an old halyard. This was the most exciting drama during our stay on Niue, and fortunately had a happy ending.

Bicycling along the road we found a delightful, tiny Italian restaurant, run by a young Italian couple. A few years before they had come to the island and fallen in love with it. They opened a one-table restaurant that serves delicious dishes such as home-style lasagna, spaghetti, and seafood linguini. It was a great way to enjoy a home-cooked lunch, at a cost of only $5. Another delightful spot along the road was the Niue Hotel, which housed the Niue Yacht Club. Many cruisers gathered at the Yacht Club bar for beers before returning to their boats in the afternoon. The yacht club commodore, Kevin, an ex-pat from New Zealand, was a goldmine of information about the island. He managed the hotel as well as the yacht club and enjoyed conversing with the many cruisers who visit the island every year. He explained the party we had witnessed at one of the houses we passed on our bike ride. It was an ear-piercing ceremony for a young girl who had come of age. All her friends and their parents bring large amounts of money as gifts, which are then passed on to the next girl on the island who comes of age. When boys come of age they have a hair-cutting ceremony, in which they lose their long locks they have grown since babyhood.

Every afternoon at 5 pm Kevin would invite visitors to watch the feeding of the fish that congregate at the surface of the water below the yacht club. He dropped pieces of leftover bread over the cliff into the water 70 feet below, and spectators would see the mad scramble at the surface of the large, red and silver fish that have been circling around ahead of time for their daily feed. A giant round boulder, weighing several tons, sits on the grounds of the hotel. Kevin told us it was thrown up by huge breaking seas when cyclone Ofa hit the island five years previously. It was an awesome reminder of the force of these winds, whipping up seas that could hurl such a boulder up over the cliff from the ocean 70 feet below. We certainly did not want to stay here until December and risk being caught in one of those devastating cyclones. Every Sunday night Kevin put on a barbecue of lamb, beef, and fresh-caught fish, vegetables, salads, and desserts for cruisers and the airport construction workers who were staying at the hotel. Kevin said the cruisers always out-compete the construction workers for food, and warned us to come early before it was all gone. It was a wonderful meal, with all we could eat for $12. We certainly got our money's worth.

In the next few days two boats we had cruised with in French Polynesia arrived at the anchorage—*Mirage*, from Portland, with Jeff and Cathy aboard, and *Asia*, from Seattle with Greg, Joanne, and their teenage daughter Hanna aboard. We decided to rent a jeep for the day to see some of the sights on the other side of the island that were too far away to reach by bicycle. The seven of us crammed into the small Landrover and traveled across to the other side of the island to see some more of the waterfront attractions. We visited a chasm that can only be reached by climbing down a 25-foot ladder to the sandy floor. Coconut palms grow in an open room surrounded by rough coral walls. We visited a cave-like structure where we swam in sea water that enters through a small opening to form a natural swimming pool, and found a beach where we could snorkel and observe the many varied tropical fish.

We learned that the island was holding a Miss Niue contest that evening, so we stopped at a small store where we could buy potato chips, chocolate bars, and cookies to eat while we watched the show. It appeared everyone on the island had turned out for this event, people sitting in family groups on mats on the lawn behind the cultural center. We watched four contestants for Miss Niue, and six for Miss Teenage Niue, compete in evening dress contest, a native dress contest, and a talent show of singing and guitar playing. It was really an impressive show, with a background of recorded music and synchronized lights. On an island were there are no movie theaters or any other entertainment we could understand why this was so well attended.

We reluctantly left Niue in early October to continue our journey to Tonga, 200 miles away to the west, and eventually on to New Zealand. Our eight-day stay on Niue was one of the most enjoyable times we had in crossing the Pacific. Sometimes the best adventures are those that are not anticipated before hand. It reminded me of the time in my youth when my father took me and my sisters and brother to see the movie *Lily*, which none of us had heard of. We arrived after it had started, so stayed for the next showing to see the beginning to find out what we had missed, and remained entranced in our seats until we had seen the movie all the way through a second time. Niue will always stand out as one of the highlights of our trip across the Pacific, and we were glad we had made the effort to visit an island that was a bit off the beaten track.

## The Kingdom of Tonga

Our trip from Niue to Tonga was an easy three days in fairly light winds. On arrival at the customs dock at Neiafu we were boarded by customs, quarantine, and immigration. The quarantine officer took half of our tomatoes, cucumbers and green peppers. We could only surmise that the officer takes the food for his own use, rather than to protect their agriculture from disease. He assured us we could replace all of the confiscated produce at the local store.

Neiafu is the capital of the Vava'u group of islands in the north part of Tonga. More than a dozen islands are spread out in a 10- by 15- mile area. The Moorings Company is located here, and charters sailboats for a week or more. There are so many anchorages that the Moorings Charter outfit gives their customers a guidebook with 40 anchorages, each one given a number. The locals on many of the islands serve feasts in the evening to the people on boats anchored in the coves. Although somewhat expensive, at $18 per person, the feasts are very popular with the cruisers.

The day after our arrival our friends Greg, Joann and Hanna, from the Seattle boat *Asia*, and Raymer, Fay, and B.R. from *Buena Fe* invited us to join them for a hike up Mt.Talau. This is actually a 400-foot high hill to the west of the anchorage. Charlie, a Peace Corps volunteer from San Diego, was our guide. He was trying (and would succeed) to obtain government approval to turn Mt. Talau into a national park. On the way to the mountain we passed a large house on the hillside with a spectacular view. Charlie explained that this was the palace where the King and Queen stay when they visit Neiafu. The trail passed through woods to the flat top where we had wonderful views of the village of Neiafu on one side, and outlying anchorages on the other side. We learned from our guide that all the feasts had been cancelled this season due to an outbreak of typhoid fever. There were 45 cases of typhoid fever in the Vava'u islands, mostly in the outlying villages with pit latrines. Flies from the latrines spread the bacteria to food, and people contract typhoid from consuming the contaminated food.

The next morning we went to the Free Methodist Church for the Sunday service. Although we didn't understand the sermon, which was in Tongan, we did enjoy the exuberant singing. At the front of the church were two huge carved wooden thrones. We were told these are reserved for the King and Queen when they visit the church in Neiafu.

After a short sail to the island of Kenutu, we enjoyed snorkeling in a coral cavern. As we drifted along the sides of the cavern we saw myriads of beautiful coral and colorful fish. There were 10 other boats at anchor, and we had a large potluck dinner ashore at the Berlin Bar. The next island, Nuku, had the best coral gardens we've seen anywhere in the Pacific. We snorkeled on the side of nearby Ava Island every afternoon for five days, and saw an abundance of butterfly fish, wrasses, damselfish, and parrotfish. We spent many evenings visiting with other cruisers on various boats at the anchorage.

After two weeks in Vava'u we departed for an overnight sail to the Ha'apai group of islands, 100 miles south of the Vava'u group. We anchored on Foa Island in front of Sandy Beach Resort. A German couple was building a resort on a beautiful white sand beach. It wasn't yet finished, but they did have a nice restaurant. From the patio they had views of whales and porpoises and a tall volcano in the distance. For several days we joined other cruisers for coffee in the morning, $5 hamburgers for lunch, and wine in the afternoon. One afternoon we took a bus over a bumpy road to town, where we stocked up with bread, crackers, bananas, tomatoes, and onions at several small markets. We visited Uoleva next, where the snorkeling on a long, coral reef parallel with the beach was wonderful. Ha'afeva anchorage was our next stop only 18 miles away. We walked on a footpath to see the cemetery, passing papaya, banana, and taro fields. Every day the local women came alongside the boats at anchor with canoes laden with fruit and their beautiful hand-woven baskets to sell to the cruisers. We visited the school, and were given baskets by the children that they had made in their craft classes. The teacher told us that the Americans don't send money to Tonga, but they do send their most valuable resources, people. She knew several of the Peace Corps volunteers.

Our visit to Tonga seemed idyllic, with the beautiful beaches, wonderful snorkeling, and the many gatherings with other cruisers for meals on board or at inexpensive restaurants. However, there was another side to Tonga that we found disturbing. On the way back from the hike to Mt. Talau we stopped by the house of an elderly couple who wanted us to look at their mats and shells. Their small house had a dirt floor, covered by mats woven from pandanus leaves. Selling woven mats and shells to visitors seemed a hard way to make a living. Every house we saw on these islands had dozens of pigs running around freely in their yards.

When we visited Kenutu, we were told by Doris, on *Balomar*, that a young woman she had visited earlier needed milk for her baby. We went with two other cruising couples and brought her canned, long-life, and powdered milk. This would supply her for a few weeks, but we wondered how her baby would fare when that was used up.

One of the cruisers, a retired nurse, visited a clinic in a village in Vava'u. She was shocked to learn that the only supplies they had were bandages. There were no antibiotics or other medications for basic medical care.

At a village in the Ha'apai group we learned that some of the villagers had gone to the United States to work. One of them sent toilets to everyone in his village. The villagers installed the toilets in concrete brick outhouses near their homes, but they were never used, because there is no running water or sewage system.

We realized that Tonga is not a tropical paradise for most Tongans. The Royal family lives in spacious palaces with every modern convenience. They fly to New Zealand for medical care. By comparison the islanders live in primitive houses, with poor sanitation, inadequate food and medical care, and try to eke out a living selling fruit, mats, and baskets to tourists and visiting cruisers. We were appalled that the Royal family has such wealth, while the rest of the people are living in such poverty. Tonga was the only country in the South Pacific we found depressing. We had no desire to return to it for another visit.

## Knockdown

Our last stop in Tonga was the capital, Nukualofa, on the island of

Tongatapu. From there we would make the passage to Opua, New Zealand. This was the final leg of our six-month trip from Mexico to New Zealand. The distance is 1,040 miles so we expected it to take about 10 days. The weather during the second week we were in Nukualofa was rough—lots of wind and rain. While waiting for the weather to improve we filled our diesel and water tanks, and bought fresh provisions. A group of 10 or 12 boats were in the same situation, and we had frequent gatherings ashore with friends, eating at the local restaurants, or sharing dinner on each other's boats. Day after day on the morning net Arnold gave us reports of gale-force winds and large seas. In the middle of November Arnold finally gave a favorable forecast for making a passage. This gave us time to make it to the safe haven of New Zealand before the cyclone season began. All of us did our last-minute preparations and departed from Tonga in the following couple of days.

The weather was much improved over what we had experienced while waiting in Tongatapu. For the first four days the sailing was pleasant, with sunny days, light to moderate winds, and slight seas. On the fifth day the winds became so light that we had to raise the spinnaker to make any progress. The wind died later in the afternoon, and we had to motor for the rest of the day. On the sixth day the weather became unsettled, with frequent squalls and rain. Each time a squall hit we had to roll in the jib, and roll it out again after it passed. The squalls continued all afternoon, but they were of short duration. On the seventh day moderate winds sprang up, and we again had perfect sailing conditions. We were making good progress towards New Zealand.

Everything changed on the eighth day. The winds increased to 20 to 25 knots in the morning, and the seas gradually built up. Heavy rain blew on to our stern nearly horizontally, soaking the ladder and the chart table down below every time we slid the hatch open to look around for ships. In the evening the winds increased further and the waves built up to enormous size. Dan put in a second reef in the mainsail, and dropped the staysail. Still we were sailing at six to seven knots, riding the huge seas like a rollercoaster. The wind vane steering kept the boat on course, but it was a wild ride. I tried to sleep on my off watch but the horrific motion and the noise made that impossible.

In the middle of the night we had heard Nedra, on the sailboat *Magic Carpet*, call a ship on VHF that they had sighted on radar. She gave their position and course to warn the ship of their presence and avoid a collision. She announced that there were at least a dozen sailboats in the vicinity, heading to New Zealand. Then came a rude reply from an unidentified boat: "What the hell are these sailboats doing out here in these conditions?" It was then we realized how dangerous this passage was. Without radar we couldn't detect ships until they were less than a mile from us, and at their speed of 20 knots, it meant we could collide within three minutes of sighting a ship.

When I took over the watch from Dan at 3 am I looked at Dan's entries in the logbook. Starting at midnight he recorded winds of 35 to 40 knots. We were in a gale. During my watch the winds continued to increase and the waves kept increasing in height. Every 10 minutes I had to slide the hatch open enough to pop my head up into the cockpit and look around for ships. It was then that I became alarmed. Our boat was rising up and surfing down enormous waves, and the wind was howling in the rigging as it gusted to 40 knots. Heavy rain was pelting down nearly horizontally. In the pitch blackness I was unable to see anything. I was frightened by the lack of visibility and the horrible motion of the boat as we surged down wave after wave, sometimes reaching a speed of seven or eight knots, seemingly out of control. I wanted to slow the boat down.

About 5:30 am, a half hour before the end of my three-hour watch, I noticed Dan was awake in his bunk when I went down below after looking around. I suggested that we heave to, which would slow the boat down to about one knot and keep it under better control. Dan persuaded me this would not be wise, as we were in the shipping lanes for Auckland, and would be sitting ducks for freighters transiting this area. I had to agree—it was better to continue on at our present speed to get through the shipping lanes as fast as possible. At our present speed we should make landfall in a few hours.

Just as we ended our discussion we were startled to hear a loud roar through the water, sounding like a freight train approaching our boat from the port side. The noise was ominous, much louder than the water rushing past the hull and the wind whistling in the rigging. We had no idea what was coming. Then it hit, with tremendous force, causing the boat to lurch forward and roll over on its side. After the initial shock we realized that we had been hit by a rogue wave. Tons of water crashed down on the cabin and decks; it was like being under a huge waterfall. The force of the wave threw me across the boat, one knee slamming against the side of Dan's berth. Dan was in his bunk on the low side, and merely rolled against the side of the hull. The boat remained heeled over at an angle of about 60 degrees for only a few minutes, then slowly, slowly came upright. It was a great relief that the boat righted itself—we could have been rolled over 360 degrees and been dismasted. Once upright the boat continued to sail down wind, rising up and screaming down the huge waves, the wind vane steering the boat.

We scanned the inside of the cabin, where several inches of water had accumulated on the low side of the boat next to Dan's bunk. The unwashed dishes had been tossed out of the sink on the port side and thrown across the cabin. Plastic bowls and pots and pans were strewn around the floor, and knives, forks, spoons and a spatula were perched above the ports and in the rail on the starboard side. The three floor hatches that cover the bilges were lying haphazardly on the floor, having been lifted up and shifted forward and to starboard. Both of us immediately began replacing the floor hatches, retrieving the utensils and placing them back in the sink, bailing the water with pans and dumping it into the sink, then sponging up the last of the saltwater on the floor. The boat seemed to be intact, no holes, cracks, or broken ports. The seawater must have squirted through the edges of the main hatch and through the open sink drain. In the forepeak the large, heavy floor hatch had been displaced and some of our clothes had fallen off the shelf onto the floor, but we quickly replaced the floor hatch and tossed the clothes back behind the netting. Everything else seemed to be in place; none of the cabinet doors had opened up to spew out their contents, and the books remained on their shelves held in by rails. We were badly shaken up, but the inside of the boat now looked as though nothing had happened.

It was now 6 am, the start of Dan's watch, so he stuck his head out to look around for ships, and immediately declared that we had a problem. Our four-inch diameter aluminum boom had snapped in two. It was bent at an angle of 45 degrees, with the foot of the main sail still in the slot. Rather than risk ripping the mainsail we decided to drop the sail and motor the last 30 miles to our waypoint. Dan had the extremely difficult job of pulling the mainsail down and lashing the broken boom to the side deck. Morning had arrived. In the light of day I was terrified by the towering 15-foot waves rising relentlessly at our stern. Dan started the engine and took over the watch, remaining in the cockpit in his full foul weather gear, watching through the gloom and sheets of rain for ships. Shortly after we began motoring Dan sighted the first ship we had seen on the whole passage, a huge container ship. Dan called to the ship on the VHF radio and talked to the captain who gave him the ship's course and position. Fortunately we were not on a collision course. I was too frightened to climb out into the cockpit to stand my watch, so for the next five hours I stayed below, with all the hatch boards in, unable to see Dan or the huge waves, but knowing our boat was in Dan's capable hands. I was shivering, my teeth chattering from fear. I worried that another wave would overwhelm the boat, or that another ship would suddenly appear out of nowhere on a collision course.

Finally, at 10:30 am, the waypoint alarm on the GPS started beeping—we had reached our first waypoint, at the entrance to Veronica Channel. We searched for the headlands at the entrance to the channel, but could see nothing except giant waves and driving rain. What a disappointment! No joyful shouts of "Land Ho!" The headlands were less than two miles away, but no land was visible.

Now we had to make a 17-mile journey through Veronica Channel before we reached Opua Harbor. The deepest part of the channel is marked by buoys spaced at a distance so that as soon as you passed one the next one would be visible. Navigating this channel should have been easy, but in these conditions it became an ordeal. I stayed below at the chart table. Before we reached each buoy I hurriedly determined the latitude and longitude of it, entered a waypoint name and the coordinates of the buoy into the GPS, and then called out to Dan the direction to head towards the buoy. The buoys weren't visible until we were about 200 feet from them. As soon as Dan was able to make out a buoy he slid open the hatch an inch or two to tell me he had sighted it. I then had to quickly enter the coordinates of the next buoy and give Dan the course to steer to it. The positions had to be precise—entering a wrong number would make it impossible to find the buoy. Our chart, which was a copy of an original chart, was sopping wet from all the rain that poured onto the chart table every time we opened the hatch, making it a difficult chore to mark the positions in pencil without tearing the paper. This process continued for the next four hours. The waves slowly decreased as we traversed the channel, but the wind and rain continued unabated. It was a miserable trip.

By the time we reached Opua at 2:30 in the afternoon we were both exhausted and traumatized by our experience. Neither of us had had more than a few hours of sleep the night before and we had eaten very little. We were tremendously happy to see the masts of dozens of sailboats scattered around the harbor and tied up at the docks—we had made it. We pulled into the customs dock and waited for the quarantine and immigration officers to come on board. Our ordeal was over. On the last day of our passage we'd gone through a gale with huge seas, been drenched by rain, and suffered a knockdown by a rogue wave. Thanks to *Shaula* being such a strong and well-built boat, and Dan having the stamina and nerves to stand a continuous watch for more than eight hours in appalling conditions, we had survived. We came through it with all intact except for a broken boom.

We stayed on the boat while the quarantine officers came on board and took all our remaining fresh food—onions, potatoes, garlic, eggs, cabbage, popcorn, dried beans, green bananas and honey. After our ordeal we didn't care. They checked our passports, and had us fill out the paperwork for immigration. By this time Customs was closed.

Gail and Bobby from the sailboat *Tulum III* came down to the dock and invited us and Ramer and Fay from *Buena Fe* to the yacht club. Gail and Bobby had arrived a week earlier with strong headwinds, making it extremely difficult to reach port. When they saw us arrive at the dock they persuaded the cook to prepare a full dinner for us at 3:30 in the afternoon, when the yacht club was normally closed. We were still in a state of shock from our knockdown and groggy from lack of sleep. Not having eaten much of anything for the last 24 hours, we enjoyed a fantastic dinner of a large steak, potato salad, green salad and orange juice. After dinner we completed our check-in with customs and immigration, and collapsed in bed at 6:30, with the intention of waking around 10 pm to have a snack. We both slept until 7:30 the next morning. When we woke up it was Thanksgiving in the U.S, but we had no energy or desire to celebrate the holiday with a big turkey dinner.

## Recovering from the Trauma

After 13 hours of much-needed sleep we began the many jobs to restore our boat to normalcy. Along with a dozen other cruisers who had recently arrived, we did our laundry at the Opua Yacht Club. Within walking distance of where we were moored was a small boatyard, owned by Doug, a fellow Bristol Channel Cutter owner. Dan took the boom over to the boatyard, in two pieces, to find out if it could be repaired. Doug suggested we take it to Whangerei, about an hour's drive down the highway, where there were all sorts of marine shops.

The huge wave that knocked our boat over managed to dump saltwater onto the quarter berth next to the companionway, and into the chart table/ice box, when the hatch on top of the icebox had swung open. To clean the quarter berth we had to take everything out and wipe it down with fresh water to get rid of the salt. The icebox, about four feet deep, also had to be emptied, and the drain on the bottom cleared so the saltwater could drain into the bilge. The dodger frame was bent as a result of the knockdown, the Sunbrella cover was torn, and the snaps ripped out. There was a canvas shop within walking distance, so we removed the cover from the frame and Dan took it to the canvas shop to be repaired.

As we began our repairs we learned that one of the sailboats that had left Tonga, the *Melinda Lee*, was overdue. The *Melinda Lee* had left Tonga a day after we left, but being a larger, faster boat it was due to arrive the same day as we arrived. We had met Mike and Judy and their two young children in Tonga in October, and had an enjoyable dinner with them at a restaurant in Vava'u along with Fay and Ramer on *Buena Fe*. *Melinda Lee* and *Buena Fe* had traveled together, and their three children were constant companions. On Halloween we were delighted to see Judy and Fay in their dinghy with seven-year-old Annie and nine-year-old Ben from *Melinda Lee* and six-year-old B.R. from *Buena Fe*, dressed in their homemade Halloween costumes, coming by our boat for trick or treat.

Two days after we arrived in Opua the New Zealand maritime authorities sent out search planes to look for the *Melinda Lee*. A plane searched all day in the vicinity of the last known position of the boat, and in the area where the boat might have drifted due to wind and current. Finally, late that afternoon, a plane that was flying along the coast reported they had seen a person next to an overturned dinghy on the beach in Deep Water Cove. It was Judy, the only survivor after *Melinda Lee* was hit by a freighter. She was picked up by a helicopter and flown to the hospital in Whangerei. The next few days we learned the details of the tragedy when Judy was interviewed by the press in her hospital bed, where she was recovering from hypothermia and head and neck injuries.

While Judy was on watch in the middle of the night, she had gone below after looking around for ships. Minutes later their boat was hit on the side where their son was sleeping, and it sank within seconds. Judy, who was on watch, and Mike and Annie, who were sleeping, were able to escape from the boat and swim to the surface as it sank. Tragically their son Ben went down with the boat. Their life raft, on deck, did not deploy. Their partially inflated dinghy was the only thing floating, having freed itself from the lashings to the foredeck.

Judy, Mike, and Annie clung to the side of the overturned dinghy through the night. Annie lost her grip and drifted away from the dinghy repeatedly, each time being brought back by Mike. After many hours of drifting in the 15-foot seas, suffering hyperthermia from immersion in the cold water, Mike and Annie were unable to hold on to the dinghy any longer. They drifted away and Mike waved goodbye. Judy clung to the dinghy for another 36 hours before finally being washed up on the beach. If Judy hadn't been spotted by the plane two and a half days after the collision, she also would have died from exposure. We were filled with grief and horror.

A few days later the yacht club let us use their meeting room for a memorial service for Mike, Annie and Ben. Everyone was grief stricken. It seemed impossible that Judy could survive and recover from the loss of her entire family. It was particularly hard on Fay, whose son had lost his two best friends. My memory of the memorial service was of seeing Fay trying to suppress her tears as she clutched an infant, borrowed for the occasion from a friend. When one of the cruisers played a beautiful rendition of "Ave Maria" on the piano I couldn't refrain from sobbing.

Cruisers, always practical, held a three-hour safety meeting that afternoon to discuss how a collision could be avoided, attended by about 50 sailors. Everyone realized that any of us could have been hit by a freighter in the conditions we experienced entering New Zealand. When approaching New Zealand ships are everywhere. There are no shipping lanes for the numerous ships departing and entering various harbors. It was scary to realize how vulnerable we were. Although a collision with a freighter is extremely rare, we now had that danger permanently in our thoughts. Having all the best safety equipment and standing watches 24 hours a day doesn't guarantee safety. After our experience I was much more frightened of sailing in big seas. We hoped we would never again encounter conditions like that in a place where there was shipping traffic. We decided to install radar on our mast. In conditions of low visibility, we also decided we would stand a continuous watch, rather than looking around only every 10 minutes. However, although this experience instilled new fears of passage-making it didn't deter us from continuing on our cruising adventure. After our sojourn in New Zealand we knew we would be sailing on to explore Fiji, Vanuatu, New Caledonia and Australia.

A week later we decided to go sailing with our friends to explore the many islands in the Bay of Islands. A new boom we had purchased for $300 was being worked on in Whangerei. The owner of the shop was removing end fittings from the old boom and fitting them to the new one so it would fit on our mast. Without a boom we were able to sail using the storm trysail instead of our mainsail. Our dodger was at the shop being repaired, but in the dry weather and calm seas we didn't need the dodger. We planned to have the dodger frame straightened later when we sailed to Whangerei. The boat would soon be totally repaired, but the repair to our psyches would take a lot longer. Our belief in the safety of ocean cruising was forever shattered.

## Another Rodent

Shortly after arriving in New Zealand we sailed down the coast of the North Island to Kawau Island, about 10 miles north of Auckland. We anchored in a small bay in front of Cris and Dick's house. Cris and Dick are sailors and bluegrass singers and players whom we met seven years before on our way down the coast of California. We had sailed across the Pacific to French Polynesia with them, but had parted company a year later when they sailed on south to New Zealand while we sailed north to Hawaii and back home to Seattle in

1990. They had settled in New Zealand, bought a beautiful house on the water, and kept their boat, *Chautauqua*, on a mooring in front of their house. It was delightful to visit them at their home.

One early morning when we were anchored in the calm harbor I woke up early and stuck my head out the main hatch to look around. I was shocked to see a small, brown rat staring at me with his beady eyes, not more than five inches from my face. I woke Dan by calling out "Dan, there's a rat in the cockpit!" Dan had the presence of mind to reply "Put in the hatch boards!" The rat trap we had used when we were boarded by a rat in the Tuamotus six months earlier was still under the sink, but we were determined to keep this rat from coming down below so we wouldn't have to use the trap.

Dan got up and climbed out to the cockpit with our large flashlight, replacing the hatch boards behind him. I stayed below, where I could hear Dan search in every nook and cranny of the boat from stem to stern. Starting in the cockpit, he tramped forward along the starboard side deck, checking behind the water jugs tied to the bulwarks, the diesel jugs tied to the mast, and under the staysail bag at the bow of the boat. After a few minutes I heard the pitter-patter of tiny feet on the port side deck above my head, heading to the stern of the boat. I yelled to Dan, "He's running back to the cockpit!" Dan tromped back to the stern to search the area again with his flashlight, and soon I heard the pitter-patter of little feet on the side deck again heading toward the foredeck. I yelled, "Now he's on the foredeck!" So Dan went forward once again to search the foredeck, but found nothing. By this time we were ready to give up. The sun had come up, and we were hungry for breakfast, so we stopped to eat breakfast with the hatch boards in place, before searching for the rat again.

After breakfast we both went on deck, putting in the hatch boards behind us, to try to find the rat. In broad daylight we would surely be able to find him. There are very few places to hide on deck. First Dan checked the sail bag on the storm trysail next to the mast, where we keep the sail hanked on ready to raise in case of a storm. The bag encloses the sail and is pulled together by a drawstring at the top, where the rat could have entered. Dan removed the sail bag, held it upside down over the side of the boat and shook it, but nothing happened. Next he removed the bag from the staysail at the bow of the boat. The sail bag is held closed by turn latches, but there are gaps between the latches where a small rat could enter. Dan shook the inverted bag over the side of the boat, when to our surprise out plopped the small brown rat. We watched it sink down a couple of feet, then slowly rise to the surface and start swimming across the harbor with a very graceful breaststroke. We were amazed at what a good swimmer it was, with its head held above water, leaving a small v-shaped wake behind. When it swam so far away we could barely see it we got out the binoculars, and watched it make a beeline for another sailboat anchored on the opposite side of the harbor. We were greatly relieved to see it leave our boat, but felt sorry for the owners of the other sailboat, who would likely have it as an unwelcome visitor when it climbed up their anchor chain.

When we went ashore later that morning to visit Cris and Dick and tell them about our surprise visit by a rat, Cris said "That was a Maori rat. We see lots of them around here." Rats were brought by the Maoris for food when they arrived in New Zealand from Polynesia in their canoes. Archeologists have found huge piles of their small bones, somewhat like the First Nation clamshell middens in the Pacific Northwest. According to modern day Maoris, when the rats are roasted over a fire and skinned, the meat can be eaten off the bones, and tastes like chicken. There's very little meat on the creatures, however, so it takes about 20 rats to make a meal.

Today, Maori rats are nearly extinct on the New Zealand mainland, because they've been out competed by the much larger and more aggressive Norwegian rats. They're now found mainly on the offshore islands, such as Kawau Island, which is where we encountered ours. We were thankful this rat didn't get below into the cabin, where it could have done serious damage to our boat.

## A Stone in New Zealand

In 1997 we had been in New Zealand for a year and a half, cruising along the coast of the North Island in our boat and traveling around the South Island for six weeks in a rental car. The cyclone season in the southern tropics was over, and we planned to continue our cruising across the Pacific, visiting Fiji, Vanuatu, and New Caledonia on our way to Australia. We began preparing for the cruise in mid-March, after our return from our road trip on the south island. The boat needed some maintenance. Dan painted the exterior wood, while I re-sewed the mainsail cover and the staysail cover, because the stitching had rotted in the sun. Dan installed our new ham radio and solar panels while I filled up our can locker with over a hundred cans for our six-month cruise. I put our photos in albums, and filed our income taxes. Between chores we socialized with the many cruisers at the marina who were also getting ready to leave New Zealand.

Our life raft was 12 years old by this time, and needed to be re-packed, which meant taking it to Auckland by ferry, then by taxi to the outfit that sells and services life rafts.

Fortunately there was a ferry leaving daily from Gulf Harbor Marina where we were staying to downtown Auckland. The life raft weighs over 50 pounds, so Dan lifted it out of the boat and into a cart, which we wheeled down to the ferry. When we reached Auckland Dan had to lift the life raft from the ferry and into a taxi to take it to the life raft outfit. They inflate the life raft to check for leaks, then replace the water, food and other emergency supplies inside, deflate it and carefully pack it back into its carrying case. They said they could have it done in about a week, and would send it back by courier.

The next morning, April 16$^{th}$, I had an appointment with a doctor in Orewa, which involved taking a 45-minute bus ride from the marina. I had a small growth on the back of my hand which the doctor had biopsied earlier. Fortunately it was benign. The doctor removed it using liquid nitrogen. I returned to the marina in mid-afternoon, but was intercepted halfway down the dock by a friend, Nedra, on the boat *Magic Carpet*. She told me that Dan had been taken to the hospital in Auckland by ambulance in the morning. He had suffered from acute back pain and vomiting after breakfast, and she had seen him walking down the dock past their boat, doubled over in pain. When she saw his pale face she immediately offered to take him to her nearby doctor's office in her car. Dan was feeling better by the time they got to the doctor's office, and the doctor thought he had strained a muscle from all the lifting of the heavy life raft the previous day. He recommended physical therapy, and was writing a prescription for a painkiller, when Dan had another attack of excruciating pain. To avoid passing out and falling, Dan lay down on the floor of the doctor's office, writhing in pain. The doctor said, "I think I'll re-consider my diagnosis. I'm going to send you to the hospital in Auckland." When Nedra offered to drive Dan there, the doctor said, "We don't do it that way. I'm calling an ambulance." So off Dan went for an hour trip to the public hospital in Auckland.

Nedra had obtained the phone number of the doctor at the Auckland Hospital so I could call and find out Dan's condition. It was good news. They had done an ultrasound when Dan arrived at the hospital to rule out an aortic aneurism. They found that he had a two-centimeter stone in his right kidney. The pain had subsided, but he was admitted to the urology ward, where he would stay until he could be scheduled for surgery to remove the stone. Although a kidney stone is serious, I was relieved to know that it wasn't an aortic aneurism, and he could be treated.

The next morning I took the ferry ride to Auckland and a bus to the hospital. Dan was in good spirits, chatting with five or six other patients in the ward, discussing the size of their kidney stones. Dan had the biggest stone, two centimeters. (The next largest was only one centimeter). I talked to the doctor, who said they would try to schedule surgery at 5 pm the next day. I brought Dan some reading material, and stayed to chat for a while, then went back to the boat. It was scary thinking Dan was having surgery by an unknown doctor, in a very busy public hospital, late the next day. There was nothing to do but wait for the outcome.

I called Dan the next morning, but he still didn't know whether they could fit him in for surgery. Then in mid-afternoon he was told they were too busy, but since he wasn't in any pain they would release him from the hospital. He was told to call the surgeon on Monday to see if surgery could be scheduled the following week. Dan arrived back at the marina on the ferry at 5:30 pm. Nedra and Chris invited us over for dinner on their boat, which was greatly appreciated. We were both in a state of shock over this development.

The next Tuesday we had an appointment with the surgeon at his office in Takapuna, near Auckland. Alby, a New Zealand friend, drove us down to the surgeon's office. The surgeon had a private practice in addition to working in the Auckland public hospital. Lithotripsy was not an option because New Zealand only had a small machine mounted in a van, which served both the North and South islands. It would probably take at least two sessions to break up Dan's 2 cm stone with ultrasound, and would be difficult to schedule. The doctor said he could remove the stone through keyhole surgery the next afternoon. The procedure involves making a small incision, breaking up the stone with a probe, then grabbing and removing the pieces by forceps. The recovery would be fast, and the doctor thought we would be able to sail to Fiji in May as planned. We would have to pay for the surgery in advance, and our health insurance wouldn't cover it fully because he was not a preferred provider. He said we would have to decide by noon that day whether or not to go ahead with the surgery.

After we questioned him extensively, wanting to know the worst case scenario, he offered to let us call our family physician for a second opinion. We left to find a pay phone out on the street, and were greatly relieved when we reached our doctor in Seattle immediately. He said he would consult with the urologist down the hall, and we could call back in 30 minutes. When we called him again he told us that surgery is no longer used in the U.S. because it is more invasive, and has a longer recovery time than lithotripsy. He recommended we fly back to Seattle, where lithotripsy could be scheduled right away. Although this meant we wouldn't be able to sail to Fiji this season, we were glad to have good advice for a less invasive treatment for Dan's kidney stone. When we told the surgeon that we were going back to the States for treatment, he expressed surprise that they could treat such a large stone with lithotripsy. I had a feeling he was disappointed we had opted out of his procedure—surgeons love doing surgery.

Changing our plans from sailing to Fiji to flying to Seattle involved much frenzied activity. I made plane reservations for Auckland to Los Angeles, then on to Seattle for Saturday, four days away. Not wanting to leave the boat in the water during our absence we scheduled the boat to be hauled out at the marina yard the next morning. We asked the hospital to send Dan's records and X-rays to us at the marina. To get permission for our boat to remain in New

Zealand for another year we had to send a copy of the medical records to customs. I called my sister in Seattle to tell her about Dan's condition and that we were arriving there in four days. Dan called his parents in California to tell them we would be visiting them after Dan recovered from his treatment in Seattle. Our life raft arrived back from Auckland and friends helped us haul it up to the boat and put it away. We got the boat shipshape, gave away our fresh food, and moved to Stan and Marj's house to stay in their spare room for the next few days. Living on a boat when it is in the yard is very uncomfortable, involving climbing a tall ladder to get on and off the boat. We had met Stan and Marj a year earlier when we were staying at Gulf Harbor Marina, near their home. They had made us feel very welcome in their home, and we had taken them for short day sails on our boat, a new experience for them. Marj and Stan served us wonderful meals, took us on drives around the area, and took us back to the boat on Saturday morning so we could pack for our trip. They then drove us to the airport in Auckland. The doctor prescribed some powerful pain pills that Dan could take if he had another attack on the plane. Previously the doctor had had a patient who flew back to the States with a kidney stone and had an attack on the plane. The plane had to divert to the nearest hospital, which was in Hawaii. The airlines tried to sue the doctor for the extra cost of the diversion. We just had to hope that Dan's kidney stone didn't shift during the flight, and if it did, that the pain would be manageable.

Fortunately, everything went well. My sister and brother-in-law met us at the plane in Seattle, and we were able to stay in their house for the summer, while they sailed to Hawaii in their boat. Dan met with the urologist and had new X-rays taken on Monday morning, a pre-op appointment the next day, and lithotripsy on Wednesday. I was able to borrow my niece's car to take Dan to Virginia Mason Hospital and Medical Center for the treatment. The technician focused shockwaves on the stone for 45 minutes while Dan sat in a bathtub of water. The technician watched the stone break into pieces on the fluoroscope until it was no longer visible. Dan had a stent inserted in the ureter between his kidney and his bladder, so the small particles could pass through smoothly. In a couple of weeks the stent was removed and Dan was back to normal, just a little out of condition from lack of exercise. Lithotripsy seemed a much easier procedure than surgery, and we were certainly glad that we had had that option. We didn't mind postponing our trip to Fiji for a year.

Although a kidney stone that shifts and blocks a ureter can cause intense pain, Dan was fortunate in having only two episodes of severe pain, both of short duration. If the stone had blocked the urine flow for a prolonged period his kidney might have shut down, and ceased to function. It was lucky that Dan's kidney stone problem occurred when we were near a large hospital, and we were able to get a diagnosis and treatment fairly quickly. It would have been traumatic to have this experience while in the middle of the ocean on the way to Fiji. We were grateful for the help of our friends in New Zealand and my sister and brother-in-law in Seattle. Having an excellent doctor and hospital facilities in Seattle made the treatment easy. The doctor said that the stone had probably been developing over many years, and as Dan was 61 when it became a problem, he would probably not have to worry about another stone for another 61 years. However, he advised Dan to drink plenty of water.

## Anchoring at Great Barrier Island

In January 1998 we sailed to Waiheke Island and Great Barrier Island, off the coast of New Zealand north of Auckland. This was our third season of cruising in New Zealand because our departure was delayed for a year by Dan's kidney stone. Our friends Brian and Sue on the boat *Waitoa* accompanied us on this excursion. Brian and Sue were from Auckland, and had cruised extensively around the area for 20 years. They knew where the best anchorages were, and where to catch fish, and collect shellfish. They left the marina before we did, but we joined them at anchorages on Waiheke Island and on Waimata Island. Brian and Sue picked oysters off the rocks, and shared them with us, and Brian caught three snappers, and gave us one for breakfast the next day. From Waimata Island we sailed to Whangaparapara on Great Barrier Island, a beautiful, unspoiled island just a day sail from the mainland. The next day another New Zealand boat, *Nerissa*, with Trevor and Raewyn aboard, sailed into the bay and dropped their anchor near us. Brian and Sue invited all four of us over for happy hour, and we were delighted to meet these new cruisers. Trevor and Raewyn had owned a kiwi fruit farm on the South Island of New Zealand, and later a deer farm. They had recently sold the deer farm and bought a rental property and their sailboat, with the intention of circumnavigating in her. It was a very well-equipped boat for offshore sailing. This was the beginning of their cruise.

The next day all three boats left Whangaparapara and headed to Port Fitzroy. *Waitoa* stopped at Greg Island along the way and picked up enough green-lipped mussels for all of us. *Nerissa* led the way, and found a small cove to anchor in with the right depth, and we and *Waitoa* followed them into the cove. There was another boat at anchor, *Greeka*. In the evening Brian invited us and the crew of *Nerissa* and *Greeka* to go ashore for a dinner of mussels on the beach. Green-lipped mussels are unique to New Zealand and are delicious. We had each cooked our mussels in different ways. Brian had smoked theirs in a small stainless steel smoker using wood chips. We had a wonderful mussel feast ashore.

The next morning it began to blow, with gusts to 40 knots, making the anchorage extremely uncomfortable. Raewyn and Trevor invited us all over to their boat for lunch, but we had to decline, because it was too risky to leave the boat in case we started to drag anchor. The gusts of wind were coming alternately from two directions, from the bow, and then from the starboard side, creating wild gyrations of the boat. In early afternoon Brian and Sue decided to raise anchor and move to a small cove nearby that is recommended in the guidebook as a good anchorage. The rest of us stayed on board our boats waiting it out, hoping the winds would diminish. Our boat was rolling from side to side, and at one point heeled over so far that it rolled over the dinghy that was tied along side. This trapped the dinghy underneath, caused the dinghy to tip and half fill with water. Dan had to bail it out to keep it from sinking. He was looking at the wild rocking of *Greeka*, anchored in front of us, when a strong gust hit the side of their boat where their inflatable dinghy was tied. Their dinghy flew up in the air, flipped over and landed on the top of their cabin, an amazing spectacle! It was a very convenient way to get their dinghy on board. Dan decided we should raise our dinghy up to prevent it from shipping water repeatedly, so we went out on deck in the 40-knot winds, and with great difficulty cranked it up. Dan held it over the cabin top while I lowered it. Now we just had the discomfort of the boat rolling side to side, which continued all night. We had to put up our lee cloths in order to sleep, as though we were on a passage. I found it impossible to sleep, but read a Patrick O'Brien novel off and on all night. We listened to the weather forecast, and learned that the winds were predicted to decrease to 25 knots the next day. However, the next morning the winds were just as strong. At 5 am we heard a loud crash and the tinkling of broken glass. Our small thermos jug, which we stored in a cabinet under the sink, had been thrown against the latch of the cabinet, opening the door, and falling out onto the floor, the glass breaking into a thousand pieces. Dan had put a spring line on the anchor chain, which stretches and lessens the jerking motion on the chain. He had to replace the spring lines twice during our stay at the anchorage when the line chafed through. I had breakfast in bed with the lee cloth still up keeping me from falling out onto the cabin sole. Trevor and Raewyn told us they had stood anchor watch from midnight to 4 am. The wind diminished somewhat so they went to bed. An hour later they collided with *Greeka*, luckily with no damage to either boat. This was shocking weather in a ridiculous anchorage.

After lunch, when the winds were still blowing 35 to 40 knots, we called Brian and Sue on *Waitoa*. They said their anchorage, in the small cove adjacent to ours, was fairly comfortable. The wind was blowing hard, but coming on the bow only. They said a large power boat had left, and now there was plenty of room if we wanted to come in. Both the crew on *Nerissa* and we raised anchor and moved over into the cove and anchored next to *Waitoa*. It was a great relief to be in calm waters again where we could have a peaceful night's sleep.

What did we learn from this experience? It was a mistake to choose a cove in which to anchor without following the advice of a cruising guide. When entering a cove in light winds it may appear to be fine, but the true test is what happens in strong winds. We learned this lesson the hard way. Coastal cruising is wonderful most of the time, but sometimes can be extremely uncomfortable and challenging.

## Collision in Harbor

During our third season cruising in New Zealand we had unusually good weather—balmy, with day after day of sun and light breezes. Our New Zealand friends said they hadn't seen such nice weather in more than 20 years. It was like the weather they remembered when they were growing up. We had experienced lots of strong winds and rain the previous two summers in New Zealand, and were enjoying the pleasant change. In fact today it was exceptionally hot—about 95 degrees—with the sun beating down on us relentlessly. It was the start of a planned month-long cruise along the coast of New Zealand's North Island. Our first stop after leaving Gulf Harbor Marina was Oneroa Bay on Waiheke Island. This is a large bay just a few hours sail from Auckland. The shore was lined with dozens of boats, mostly New Zealanders. We had stayed overnight, and were hoping to leave during the morning for a short sail to the next anchorage at Waimate Island. When we woke there was no wind, so we whiled away the time, hoping the wind would pick up so we could sail rather than motor. It remained windless all morning, so we resigned ourselves to motoring to the next anchorage. It was so hot in the sun that we left the cockpit awning up, which is against our usual practice, because it meant we couldn't stand up and see over the top of the dodger and dinghy. The only way to see directly forward was to step to the side of the cockpit and lean out to see around the dodger. I went forward and raised the anchor and returned to the cockpit to motor out of the bay. Dan was up on the bow stowing the anchor on deck as I began motoring. I had checked before we left and it appeared we had a clear path out of the bay. Suddenly I heard a loud clunk and the boat came to an abrupt stop with the motor still running full speed. I thought we had hit a large log, but couldn't see anything ahead of us. Dan looked up from his position on the foredeck and yelled, "Take the engine out of gear." I quickly put the gear in neutral and was stunned to see we had hit a small New Zealand sailboat anchored near the entrance of the bay. Our bowsprit, which sticks out seven feet in front of our boat, had struck the smaller boat's pulpit, lifted it off the deck, and pulled the foredeck up with it. There were four gaping holes in the plywood deck where the base of the pulpit had been bolted down. It was a shock to see the damage we had caused. The owners came up on deck to see what had happened. They were badly shaken up hearing the crash and the ripping of plywood while at anchor when they least expected it.

Dan took over the helm and we circled around to re-anchor near the boat we had hit. We rowed over to the boat and met the owners, Jim and Denise. Jim said his wife was still shaking. Their pulpit was badly mangled, and the first two feet of their foredeck had been pulled away from the hull. I apologized repeatedly, saying it was entirely my fault, and we would pay for all repairs. We didn't carry insurance while cruising offshore, because it is prohibitively expensive. Now we had to pay out of pocket for whatever damage we had caused. Dan got a tube of sealant from our boat and helped Jim pound the torn part of the plywood deck with a hammer and re-attach it to the hull using the sealant. They removed the mangled pulpit and put it in Jim's dinghy. This quick repair would enable Jim and Denise to return to Auckland where they could have the boat hauled out and repaired properly. We exchanged names and addresses, and I promised we would contact them in about a month when we returned from our cruise down the coast to Tauranga. This was a horrible start to our cruise. I felt really sorry that I had caused such havoc to this nice New Zealand couple. It was an expensive mistake, but worse than the expense was the regret for being so careless as to collide with a boat at anchor.

After returning to our boat and having lunch, we raised anchor again and motored to Waimate Island, this time with the cockpit awning rolled up and not interfering with our vision. We had four wonderful weeks cruising along this part of the New Zealand coast we had never seen before. After leaving Waimate Island we sailed to Mercury Island and joined our friends Brian and Sue on *Waitoa*. Mercury Island is a seldom-visited offshore island with beautiful white sand beaches, and several small coves for anchoring. Brian and Sue joined us for swims in the crystal clear water, snorkeling, walking on the beach, and hiking across the island and to the top of a ridge with a panoramic view of the area. After a week Brian and Sue returned to the Gulf Harbor Marina, while we continued to enjoy the many small coves on Mercury Island. Finally we motored and sailed on down to Tauranga, catching a large tuna on the way.

At Tauranga we pulled into a marina, and were greeted by a local couple, who were friends of Mavis and Alby. Mavis and Alby were cruisers we had met in Gulf Harbor Marina, who led a very active life after retiring from teaching. Their friends, Brian and Liz, took us to see the local attractions, including Mount Manganui and the hot springs, where we swam in a natural thermal pool, like a giant hot tub. A couple of days later Mavis and Alby picked us up and took us to their home near Lake Taupo. They took us on a hike along the Tangiriro River, to a fish hatchery near Lake Taupo, a hike around Lake Rotopounami, and a swim in Lake Taupo. This was a beautiful area, and our stay with them was one of the highlights of our visit to New Zealand.

The day after we returned to Tauranga from Mavis and Alby's home we stayed on our boat because of the awful weather—a 40-knot gale. This didn't stop the local Tauranga Yacht Club from sailing in a regatta out in the bay. It was a fundraising event, where local business people donated $125 to charity for the privilege of sailing in a boat race with local boat owners. We were amazed they would even leave the marina in such conditions, especially with completely inexperienced crew. We listened in awe to the stories of the returning sailors—several sailboats were knocked over, one broke its boom, and three crew members, all non-sailors, fell overboard. (Fortunately they were quickly retrieved). Our friends, Trevor and Raewyn, who had taken Coast Guard members out on their boat, *Nerissa*, thought it was great fun. Most New Zealand sailors think nothing of sailing in 30-to-40-knot winds, which occur frequently in their waters. We thought it was crazy and dangerous. On our last day at the Tauranga

Marina we went to a large potluck dinner at the marina, where we met many New Zealand and a few Canadian and American cruisers. Then it was time to head back to Gulf Harbor Marina, stopping at Mercury Island and Oneroa Bay on the way.

During our cruise along the coast to Tauranga the problem of paying for the damage our collision at Waimate Island was always in the back of our minds. Back at Gulf Harbor marina I called Jim to find out what we owed for damages. He had received several estimates for the repairs. Over the next few weeks we exchanged letters, faxes and phone calls about the costs of the repairs. Jim was reluctant to report the accident to his insurance company for fear his premiums would increase. Since I took full responsibility for the accident I didn't think he had to worry. It was not until six weeks later, after convincing Jim to use his boat insurance to coordinate the repairs, that I got the final bill from his boat insurance company. It was $1,200 (about $600 U.S.) I promptly went to the bank to get a cashier's check for the amount and sent it off to the insurance company. It was a tremendous relief to get it over with. Now we could concentrate on getting ready for our next adventure. We would be sailing to Fiji, then on to Vanuatu, New Caledonia and Australia in about six weeks. My older sister, Diane, helped me put the whole incident in perspective by saying, "It's only money—at least you weren't injured or killed." I stopped agonizing over my mistake. The collision had been an expensive learning experience. We would never again attempt to sail or motor with the cockpit awning blocking our vision.

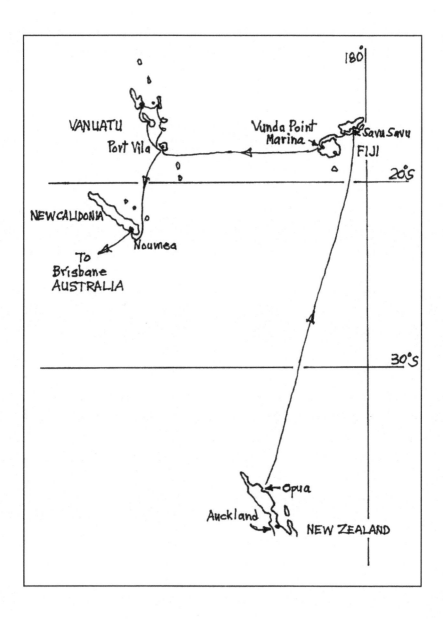

*Shaula's* passage—New Zealand to Australia

# NEW ZEALAND TO AUSTRALIA

## Savusavu

Our stay in New Zealand was coming to an end after two and a half years. We wanted to be in Australia before the hurricane season began in November. On the way we planned to explore the islands of Fiji, Vanuatu, and New Caledonia.

We left New Zealand at the end of May, equivalent to the end of November in the northern hemisphere, or late fall. Because of the chilly weather we wore foul weather gear over our warm clothing. The winds were moderate—we had waited until a low passed through the region before departing. However, the seas were horrific. The ocean had been churned up by strong winds from all directions, making the seas wild and unpredictable. Because of frequent squalls, we were cold, wet, and miserable. As we pummeled through the seas our boat was thrown around as though in a washing machine, waves hitting first on the side, then the bow, or the stern, often breaking into the cockpit and soaking us. It was the roughest and most uncomfortable passage we had ever made. It was only in the last three days of the passage, as we approached Fiji, that the weather improved. Now the breezes were warm. We began to see tropicbirds, gannets, and flying fish all around us. We stripped off our foul weather gear and were down to shorts, shirts, and bare feet. However, the seas were still rough, and when waves broke into the cockpit they soaked us to the skin.

Our destination was Savusavu, a small village on the south end of Vanua Levu, which is the second largest island in Fiji. Several other boats left Opua when we did, but they were headed to Suva, so we were completely on our own. The waters around Fiji are dotted with coral heads, so we had to be constantly vigilant as we approached land. During the last day one of us steered while the other sat watch on the foredeck. Finally, at noon on the eleventh day after leaving New Zealand, we entered the pass into the creek where the village of Savusavu is situated. We anchored in about 30 feet near a line of six boats tied to mooring buoys. Savusavu Bay was wonderfully calm and quiet. Exhausted from the rough passage, we had lunch, and immediately took naps. We woke to the alarm at 3 pm to check into the maritime net to tell them we had arrived safely, had dinner, then fell asleep again, and slept through the night.

The next morning Simon, the marina manager, called the health and agricultural inspectors to tell them we had arrived. Dan rowed the dinghy to the dock to pick up the large Fijian health inspector and take him to our boat. He gave us a long form to fill out concerning our health, and whether or not the rats on board had the plague—a question we had never been asked before. Thankfully, we didn't have any rats aboard. After Dan took the health inspector back to dry land we both went ashore to check in with customs and immigration, and fill out more forms. There was a small fee, $16, to pay at the hospital for the health inspection. Our walk along the path to the hospital was our first glimpse of the beautiful tropical plants. Papaya, breadfruit, hibiscus shrubs, mango and banana trees lined the path—a lush tropical paradise.

We took showers, and had lunch at Captains Café at the Copra Shed Marina. There we met Bruce and Kris from the sailboat *Nashira*. They had admired our boat when we arrived, and we learned that they had owned a Bristol Channel Cutter but they had to sell it and buy a larger boat after their two sons were born in American Samoa. They were long-term cruisers—their sons were now eight and 10 years old. That evening Kris and Bruce came over to *Shaula* for happy hour of wine, peanuts and hot peas. We learned that they and the people from several other boats here had spent a year, including the six-month cyclone season, in Savusavu. Fortunately, it had been a mild season with no cyclones.

Other sailors we met, who had spent the previous year at Savusavu, were Jonathon and Rogers with their two children on *Sea Hawk*; and Charlie and Jeannette on *Quark*, who had been cruising for 20 years on their boat and had circumnavigated once. Jeannette had worked as a charter captain in the Caribbean and Charlie had worked for the charter company solving mechanical problems on the boats. One boat was at the dock, *El Corazon*, and we met Cris and Tim and their two-year-old and one-month-old sons. A midwife had delivered the younger son on board *El Corazon* at the dock at Savusavu. The largest boat, the 55-foot *Impossible*, was sailed by Jim Whittaker (who became the first American to summit Everest in 1963), his wife Diane and their two sons. Jim and Diane had sailed to Hawaii on a shakedown cruise, but this was their first cruise with their two sons. Diane was an excellent photographer, and one evening while we were having happy hour she couldn't resist running down to the dock to take a photo of the sunset and a perfect rainbow over their boat. She had the facilities for printing out photos on board, so returned with a beautiful souvenir to give to Simon at the marina. The children from all the boats were having a great time, with scavenger hunts, and a birthday party where they swam at the nearby Hot Springs Hotel. We enjoyed sharing cruising stories over happy hours on board, and lunches and dinners ashore.

The Copra Shed Marina was a huge old wooden building along the waterfront with a long dock in front. Along with the marina office, it had everything a person could want—restrooms, showers, laundromat, and a small restaurant. Simon was a young New Zealander, who provided moorage on the dock and on mooring buoys in the creek. Every Saturday night he had a hamburger barbeque for $6 for all the sailors at the marina. While we were there he set up a large saltwater aquarium in front of a window on the waterfront side of the building. The aquarium was stocked by divers who dove in the warm waters around the marina and collected all sorts of wonderful anemones, colorful and exotic fish, and seahorses. Soon the children were joining in and diving to bring sea life to the aquarium. They were also learning marine biology, as Simon insisted that they identify any creature they added. Every day when we went ashore we checked out the new delights the aquarium had sprouted overnight.

Savusavu turned out to be the highlight of our Fijian cruising. We were glad we had chosen it as our landfall—we had all the facilities of a marina without having to tie up to a dock and pay high moorage fees. The small village had a good grocery store for re-supplying, and a good restaurant, without having the noise, traffic of a major tourist town such as Suva. Best of all we met many cruisers who shared our cruising lifestyle and had amazing stories to tell of their adventures.

## Many Faces of Fiji

After our six-week visit to Savusavu and the nearby anchorages, we departed to see more of Fiji. Our first destination was the small island of Mokagai, a government research station. The Fijian islands have fringing reefs with a narrow pass through which boats enter. It is best to travel between 9 am and 3 pm, when the sun is high enough in the sky so that the coral reefs are visible. Traveling through the islands is nerve-wracking. One of us had to stand on the foredeck as we approached an island to watch for coral heads. Mokagai was too far a destination from Savusavu to make it in one day, so we sailed to the small island of Namena and anchored overnight. We arrived at the island of Mokagai the next day in mid-afternoon, and anchored in the calm, quiet bay.

The next day we went ashore to visit the fisheries experimental station. One of the workers, Ben, gave us a tour. They were culturing giant clams. These clams grow to two or three feet across, and have a fluted edge at the top, with beautiful blue, green and purple algae in their opening. Giant clams were harvested to extinction by native Fijians, who ate them raw, marinated in lime juice. Recently the researchers brought in 50 giant clams from Australia and put them in concrete tanks to spawn. When they grow to a half inch the baby clams are placed in large saltwater tanks and covered by screens to protect them from birds and other predators. The workers feed them on yeast and algae. After many years the giant clams will be replanted on reefs in protected areas on different Fijian islands.

Another project is saving the hawksbill turtle. After a turtle lays its eggs on the beach, the workers cover the nest with nets to keep the eggs and baby turtles from being eaten. They collect the newly hatched turtles and place them in net-covered concrete tanks. After a few days, when the baby turtles are strong enough to fend for themselves, they are released to run down to the sea. In three years these turtles will return as adults to the same beach to lay their eggs. We were really impressed by the efforts of the Fijians to preserve their endangered species. The water at Mokagai was crystal clear, and we enjoyed snorkeling in the bay near our boat to see the sea life on the coral heads. In the mornings we took walks to observe the amazingly colorful birds in the trees—red-green parrots, kingfishers, pigeons, red-throated honeyeaters, and red-breasted blackbirds.

Our next destination was the small island of Yadua. It took two days to reach Yadua sailing only in daylight. After we arrived and anchored the winds picked up and we were unable to go ashore for two days. Finally the winds calmed down and we went ashore in the dinghy to walk to the village. It was a long trail up over a ridge to the other side of the island. When we arrived three women met us and took us to Peter's house. Peter gave us a warm welcome. He was the warden and a teacher in the elementary school. His wife prepared tea for us at their home. Then Peter took us to the chief's house where we could present the chief with a bundle of kava roots. The chief sat cross-legged on a mat on the floor; Peter, Dan, and I sat opposite him. Four young men were lounging on the floor at the side of the room in a state of torpor. Dan set our bundle of kava in front of the chief, and the chief gave a short speech welcoming us to his village. Then Dan gave a speech telling the chief where we came from, and how honored we were to visit the village at Yadua. The chief told us we were free to go wherever we wanted in the village. The speeches were interspersed with clapping. The four young men roused themselves enough to join in. This was our first sevusevu ceremony. Fortunately we didn't have to drink any kava. Kava is made by grinding the kava roots in water to form a gritty brown liquid. It is a mild narcotic that has a numbing effect on the body, starting with the legs. After we left Peter told us the young men at the chief's house were known as the Morning Glory Boys. They drank kava all morning and afternoon. Peter took us around to see the rest of the village. The people lived in bures, small thatched huts made of bamboo. He took us to visit the elementary school, where the students introduced themselves and entertained us by singing.

When we went ashore late the next afternoon we saw a procession of women and children carrying woven baskets of nerites (small black shells), coconut crabs, and small fish. They were taking the long trail back to the village. We hoped they would make it before sunset. Before we left the next morning Dan went ashore to give the other five bundles of kava we had bought in Savusavu to one of the villagers we had met. We wouldn't need the kava, because we didn't have time to visit any more small villages.

# NEW ZEALAND TO AUSTRALIA

It took three daylight sails to reach our last destination, Vuda Point, on the large island of Viti Levu. We tied up at the small circular marina. It was great being back in civilization: We had pizza at The Landing Café and wine at the yacht club with fellow cruisers who had arrived before us. They told us about the restaurants and the wonderful municipal market in Lautoka. They also told us to beware of Fijian men in Lautoka, who stopped tourists on the sidewalk, asked their name, quickly wrote their name on wooden knives they had carved, and then insisted they buy them for $30 each.

The next morning we took a 50-minute bus ride into Lautoka, the second largest city in Fiji. We saw a movie, "Primary Colors", and then had lunch at Gopal's vegetarian Indian restaurant. We stopped at the ANZ Bank to get $300 cash so we could buy provisions at the market. An older Fijian man approached us on the sidewalk, introduced himself, and said he had seen us at The Landing near the marina the previous night. He related the sad story of his young son who was in the hospital with pneumonia. He needed $43.50 to buy antibiotics for him. We stopped in a small coffee shop, where he told us more about himself. His wife worked in the pharmacy just down the street. He promised to meet us later in the day at the market to repay us after he got his paycheck. Dan and I decided we should help him out, and gave him $50. As Dan was paying for the coffee, we watched the man leave, turning in the opposite direction from the pharmacy. We realized we had been taken. Oh well, live and learn. At least we hadn't bought a useless carved knife just because it had our name on it. We continued down the road to the municipal market to stock up on produce before returning by bus to the marina.

A few days later I joined a group of cruisers, Sue, Clare and Suzanne, to shop at the Cost-U-Less at Nadi, 20 miles south of Vuda Point. Sue arranged for a taxi driven by a young East Indian man, Ali. He would take us to the store, wait for us to shop, and bring us back, all for $22. The store had an amazing selection of groceries. I stocked up on many items I hadn't seen anywhere else, including grape nuts, dried apricots, wheat thins, and chocolate chips. The trip back was terrifying. Ali drove at breakneck speed, swerving in and out of traffic with several near misses. We were greatly relieved we made it back to the marina in one piece.

After a few more days at the marina, talking to other cruisers over pizza and wine, it was time to leave for Vanuatu. We took the bus to Lautoka to check out with customs, and then took a long walk to immigration at the other end of town. While in town we bought a week's worth of produce at the market, and more groceries at Morris Hedstrom, Fiji's largest wholesale and retail organization.

Although the Fiji islands were beautiful, we found sailing in the waters around Fiji dangerous due to the many coral reefs. We were always on high alert to avoid hitting coral heads. However, we were impressed by the research station and wonderful sea and bird life on Mokagai Island, and the traditional way of life in a tiny village on Yadua. We enjoyed modern life in Nadi and Lautoka on the big island of Viti Levu, where we encountered a Fijian con artist and a wild East Indian taxi driver. Fiji was certainly a country with many faces.

## Visiting Port Vila

After visiting Fiji we planned to visit two more island groups—Vanuatu and New Caledonia. In order to avoid hurricanes we wanted to reach Australia by mid-October. Visiting Fiji had taken a couple of months. This meant we would have less than two months to visit Vanuatu and New Caledonia.

We had an easy five-day sail from Musket Cove in Fiji to Port Vila,

Vanuatu, arriving August 10$^{th}$. Friends we had met in New Zealand, Stan and Marj, were on shore to meet us after we checked into the country with customs and quarantine. They offered to be our guides to Port Vila, a place they knew well and loved. They had flown to Port Vila a month earlier to visit their son Peter and family. Peter was a scuba diving instructor, and his wife Nicky was a teacher in a private school in Port Vila. They had lived and worked on Vanuatu for over five years, and their two sons were attending the private school where Nicky taught. Stan and Marj visited every winter, staying in a nearby cottage, helping Peter and family with projects around the house.

The town of Port Vila, the capital of Vanuatu, is stretched out for a couple of miles along the waterfront of the large harbor. Although it is a small town of about 35,000 people, the main road is lined with banks, several restaurants, an Internet café, a copy shop, a library, a cultural center, a pharmacy, and a large open-air market.

We kept our boat on a mooring buoy in the large, deep harbor. Every day we went ashore and walked to town to take care of business. At the small Internet café we checked our email and sent replies. The Internet service was so slow that it took five minutes or more to send a short message, and was expensive, at $1 per minute. We decided that in the future we would use regular mail which cost only $.50 per letter. Every few days we shopped at the large open-air market for produce grown by the locals in their gardens. We walked to the Australian embassy to pick up the forms to apply for an Australian visa, and spent hours filling out the necessary information. Australia is the only country we visited that required a visa to be obtained from outside the country before arrival. A few days after turning in the paperwork we were granted a two-year multiple-entry visa. This allowed us to stay in Australia for six months at a time over a period of two years, but each time we left we would have to obtain a stamp in our passports in a foreign country before being allowed back into Australia.

During our wandering through town we stopped at the Club Vanuatu, where there was a large screen TV tuned to CNN. It was the eve of President Clinton's impeachment proceedings and we watched Patrick Buchanan and Jesse Jackson comment. Buchanan said Clinton would probably be impeached for committing perjury, and Jackson said he would be forgiven by the American people, who could separate personal behavior from public policy. The next day we watched Clinton's address to the nation apologizing for misleading the American people about inappropriate relations with Monica Lewinsky, criticizing the Ken Starr investigation, and asserting his right to a private life. It seemed incongruous to be watching impeachment proceedings against our president thousands of miles away during our idyllic visit to Port Vila.

Stan and Marj were wonderful at showing us all the sights around Port Vila. Every morning after we had breakfast on the boat we met them at Rossi's, just a few blocks' walk from where we were moored. Rossi's was a large waterfront restaurant with a panoramic view of the harbor, filled with ex-pats from Australia and New Zealand. They were having a leisurely coffee, while chatting and reading the Sydney newspapers, which were spread out over long tables. We sipped coffee or lattes while discussing the plans for the day. For two weeks we were thoroughly entertained. Stan and Marj took us north of Port Vila in their car to visit the Cascades, a series of waterfalls along a wide stream. They hired a guide they knew, Moses, who led us on a beautiful hike along a wide, cascading stream. All along the shore of the stream we saw native gardens of red and white taro, manioc, kava, corn, tomatoes, and yams. At the top of the stream a spectacular waterfall fell 50 feet into a clear blue pool. It began to rain as we descended, so Moses cut off a large leaf from an elephant ear plant for each of us to hold over our heads to keep the rain off—a clever and effective disposable umbrella.

Another day, Stan and Marj drove us a few miles away from the hustle and bustle of downtown Port Vila to a peaceful, spacious resort built on the lagoon south of Port Vila. It had beautiful grassy lawns surrounded with coconut palms, a large swimming pool, and fancy restaurant with seating outside under the palm trees. At lunchtime Stan and Marj fixed us sandwiches at their apartment or we had hamburgers at the waterfront bar near our moorage. One afternoon they arranged a free sunset cruise on a catamaran owned by Nicky's sister Rosie and her partner Danny. They took us and some of their friends on a calm sail across the harbor to watch the sunset. They served enough snacks that we didn't need to eat dinner that evening. We had Stan and Marj over to our boat for dinner one evening, and before we left we took them out to dinner at the Harborside Restaurant, where we had wonderful tender beef from cattle grown on the island, plus chicken and calamari.

Our stay in Port Vila was filled with sightseeing and was a great introduction to Vanuatu. Stan and Marj made it possible to visit many places we would never have seen on our own. We wanted to see some of the other islands of Vanuatu, so we reluctantly said goodbye and continued on our way.

## New Caledonia

Our last stop before sailing to Australia was New Caledonia. This island group is midway between Fiji and Australia. Our destination was Noumea, the capital, located on the southwest side of the main island, Grand Terre.

The passage from Port Vila to Noumea takes only three days in favorable winds. At first the passage went well, with a nice following wind. The second day we went through a squall in the early morning, then torrential rain, and the most lightning we had ever seen. The lightning was far away, but was a spectacular display. We arrived in the early morning at Havannah Pass, the narrow channel in the fringing reef that surrounds the island. After negotiating the pass using waypoints we had entered into the GPS, we tied up at the customs dock at the marina at Port Moselle. The quarantine officer who came on board was a native Melanesian, known as a Kanak. He kindly allowed us to keep all our fresh food, as long as we consumed it on board. All visiting yachts are given one day's free moorage at the marina and a voucher for a free drink at a fancy bar near by. This was an unusually welcoming port for cruising sailors.

Noumea is a modern city, with paved roads, luxury cars, modern shops, restaurants, and supermarkets. Most of the population of New Caledonia is concentrated in Noumea. It is known as the Paris of the Pacific. It was in striking contrast to the more primitive culture in the small villages of Vanuatu that we had just visited. Adjacent to the marina was a large farmer's market, with local fruits and vegetables spread out on long counters, a fish market, and a small café. Every morning after breakfast we walked to the café to have our morning coffee and croissants. The coffee was served in a porcelain cup the size of a large soup bowl—no need to ask for a refill. The croissants were freshly baked, and our favorites had a dollop of chocolate in the center. During our wandering around town we bought French baguettes and pastries at a wonderful French bakery, and had delicious lunches at French and Vietnamese restaurants at very reasonable prices.

For the next week we enjoyed the tourist sites around Noumea. The Jean-Marie Tjibaou cultural center, five miles from the city center, had opened just four months previously. We took a bus ride through the countryside to the most spectacular building we had ever seen. It was designed by the famous Italian architect, Renzo Piano, and consisted of a cluster of tall circular towers made of a combination of shiny metal and wood. Inside were many rooms of Pacific Islanders' crafts, although the exhibits were sparse. When we first entered, a Kanak guide who spoke English asked us if we spoke French. He offered to give us a guided tour of a huge room of photos and information on Tjibaou, a local leader of the efforts for New Caledonia to gain independence from France. Since all of the placards were in French we appreciated having a guide. It was impressive that the French devoted an exhibit to Tjibaou. He was assassinated in 1989 by another member of the independence movement who thought he had capitulated to France in signing the treaty of 1988.

Another day we visited the aquarium, located on the south shore of the island, only a short bus ride from town. It was entirely lit by sunlight. There were aquariums of local tropical fish and a variety of corals, displaying beautiful fluorescent colors in ultraviolet light. A couple of days later we attended a small cultural fair in Dumbea, a small village northwest of Noumea, reached by a local bus. They had set up booths displaying the crafts from the Solomon Islands, Fiji, Indonesia, and Marshall Islands. In the afternoon they held an impressive program of dances from the islands of Indonesia and Martinique (similar to Tahitian dances). The day before we departed for Australia we visited the Forest Parc, a short bus ride north of the city. The bus let us off at a huge building, which housed workers at the nearby nickel refinery. It was just downwind from the refinery. From there we walked a mile up hill to the park. It was a small zoo in a park-like setting, of local birds and mammals. We seemed to be the only visitors. That evening we attended an excellent free concert of Mozart, Vivaldi and Bach at a cathedral in the heart of Noumea. In the week we visited we were thoroughly entertained. It felt like being in Europe. However, we met very few of the local Melanesian people.

Why was this island so different from all the other islands we visited in the South Pacific? The answer is nickel. New Caledonia holds 40 percent of the world's reserves of nickel. It is extracted by open-pit mining, and partially refined in the north part of the city of Noumea. Much of the countryside outside Noumea is scarred by huge open pits surrounded with mounds of red clay. The soil is red because of the iron content. The nickel is exported mainly to France and Japan, although a small amount is exported to the U.S. It is a highly lucrative business.

The population of Noumea is predominantly French. When New Caledonia became an overseas territory of France in 1946, the French assumed title to two-thirds of the land on the main island, Grand Terre. One fourth of the land was given or sold to white settlers, 10 percent to natives. Most of the arable land is now owned by about a thousand French settlers in and around Noumea. The native Melanesian population lives far away in hilly country that is less desirable. They live in small villages, where they grow yams, manioc, and taro. Mining and smelting throws up nickel dust, which is carcinogenic. The World Health Organization rates nickel dust as one of the most toxic elements on earth, with no safe level of human exposure. The workers in the mines and in the refineries suffer from the effects of breathing the dust. New Caledonia has the highest incidence of asthma-related deaths in the world. Half the population enjoys the benefits of nickel mining, while the other half suffers the consequences. For France, the mining of nickel is a bonanza, worth billions of dollars.

All efforts of the Kanak people to gain independence from France have failed. Although a vote on independence was held in 1998 it didn't pass. The French continued to bring settlers into Noumea from France, keeping the Kanak population in a minority, guaranteeing they wouldn't win the vote for independence. France is determined to keep New Caledonia for its mineral riches.

The treatment of the Kanak population, the destruction of the environment, and poisoning of the air were so appalling to us that we had no desire to return to New Caledonia after our first visit. However, because it is on the way to Vanuatu, which we wanted to visit again and again, we later stopped at Noumea when we sailed to Vanuatu. It was a place to rest and recuperate from a rough passage from Australia. We enjoyed the coffee, chocolate croissants, fresh produce, and the company of other cruisers. However, we were always eager to leave after a week or so for the short passage to Vanuatu.

# 9 AUSTRALIA

## Coastal Cruising

In October 1998 we achieved our initial goal for our extended cruise—to sail to Australia. It had taken five years, because we cruised in Mexico for a year and a half, in New Zealand for two and a half years, and visited numerous South Pacific islands on the way. We didn't make any plans beyond Australia. We would make the decision later as to where to go next.

Australia beckoned us because we had lived and worked near Sydney from 1971 to 1974. During that time we sailed up and down a short stretch of the New South Wales coast during every school holiday, and many weekends. Our sailboat, which was all we could afford at the time, was a 23-foot racing boat without standing headroom, and only an outboard motor. We were novices at cruising. However, we were enamored by the beauty of the coastal areas, the many small harbors all along that stretch of the coast, and the friendliness of the Australian cruisers with whom we shared anchorages. Now we had *Shaula*, our large, comfortable boat built for cruising, and we had years of cruising experience. We looked forward to returning to some of the places we had visited before, and exploring hundreds of miles of the New South Wales and Queensland coasts.

Before we arrived in Australia we had obtained two-year multiple-entry visas. These visas allowed us to stay in Australia for only six months at a time. After six months we had to leave, and have our passports stamped in another country before returning. Since we first entered Australia in October 1998, we had to leave by April 1999. It had been two years since we visited the U.S. so we planned to fly home in March to visit friends and family for five or six months. When the two-year multiple-entry visas expired, we had to apply for new visas, which can only be done from outside Australia. We could obtain two-year or four-year visas, depending on what Australian Immigration decided to grant. The application fee was $175, so it was best to get the longest visa we could.

October is the approach of summer in Australia, and tropical Queensland can be unbearably hot from November to February. To escape the heat we headed south to New South Wales, where the weather is more temperate. Our destination was Sydney, but we stopped at many small harbors along the way. Contrary to the easy sailing we had when we lived in Sydney in the 1970s, the sailing between Bundaberg and Sydney had many challenges. Some of our favorite places along the coast were located on rivers, which meant we had to cross a bar, and time the entrance with the tides. The timing was difficult, because there was usually a strong south-setting current, sometimes reaching four knots, along the New South Wales coast. It could sweep us past the river entrance when going south, or delay our entrance into harbor when sailing north. In addition, there were shallow, narrow passages, such as Sandy Straits south of Bundaberg and the Broadwater south of Brisbane, which required constant vigilance to avoid going aground. Nevertheless, we found visiting the many anchorages along the coast delightful. Sydney had a well-protected anchorage, Blackwattle Bay, with a city park on one side, and the Sydney Fish Market on the other side. It was a convenient anchorage from which we could visit the museums, restaurants, coffee shops, movie theaters, a supermarket, and the Opera House, all within walking distance. It was a gathering place for foreign cruisers, and we enjoyed the company of many Australian and foreign cruisers while anchored in Blackwattle Bay.

Before we flew back to the States we hauled the boat out in a boatyard that had a secure storage area, rather than leaving the boat in the water. Our first haulout was at Oyster Cove, just a hundred miles north of Sydney, but Yamba, further up the coast, became our favorite place for a haulout, with its friendly staff and spacious boatyard.

We escaped the Australian fall and winter weather to spend the spring and summer in the U.S. It was an opportunity to get off the boat and live in luxury on land for five or six months. We spent several months in Glendale, California, visiting Dan's parents. They were still living in the house that Dan and his brother grew up in. This was wonderfully spacious compared to living on our boat. We could watch television, prepare meals in a large kitchen, help out with household chores, work in the garden and use the car to take Dan's folks to grocery stores and medical appointments. We could drive to the nearby restaurants with Dan's folks for wonderful seafood or Thai meals. After a few months we took a train up to Seattle to visit friends and family and take care of our medical appointments. Our friend Dave, who had played in string trios and quartets with me for years, was living in and restoring a large, beautiful home in the Queen Anne neighborhood. He invited us to stay in his third-floor guest room while he and his crew worked on the house. It had a spectacular view of the harbor, downtown Seattle, and the space needle. Dave was a gracious host. He picked us up at the train station when we arrived from Los Angeles. He was a collector of vintage cars, and always loaned us one of his cars or vans for our stay in Seattle. After a few weeks of wonderful visits with friends, and visits to our doctor, dentist, dermatologist and optometrist, we took the train back to Glendale for a short visit with Dan's folks before flying back to Australia.

When we returned from visiting the U.S. we worked on the boat for a week or more, applying anti-fouling, painting the exterior wood if necessary, and greasing the seacocks. Then we had another summer and fall (December to June) to cruise the New South Wales coast, from Yamba south to Sydney. Before the six months was over we sailed north as far as Coffs Harbor in order to sail to Vanuatu and New Caledonia.

Coffs Harbor was the ideal harbor for departing for Vanuatu and New Caledonia for the southern hemisphere winter. We allowed plenty of time to do our last-minute preparation for the offshore passage and to wait for favorable winds. The cyclone season in that part of the world is from November to June, so we visited these islands during the winter, the non-cyclone season. Even in winter Vanuatu is hot—we wouldn't want to be there over the summer. Vanuatu was our favorite country to visit. The people were friendly, and we enjoyed getting to know them and learning about their culture. Each time we visited we planned to return in two years to some of the places we enjoyed before and some new places. We especially enjoyed the Banks Islands, in the north part of the country. The locals were always welcoming, and seemed content to live a simple life in their small villages. New Caledonia was only 300 miles southwest of Vanuatu, and was a convenient place to stop either on the way to Vanuatu or on the way back to Australia. For us it was of secondary interest to Vanuatu, and we spent only a week or 10 days there each time.

This became our pattern for our cruising in Australia—sailing up and down the East Coast for six months, then in alternate years visiting the U.S. and cruising to Vanuatu and New Caledonia. It was an interesting and varied life. The contrast was remarkable. One year we flew by jet from Australia to Los Angeles in 12 hours, spent months in luxury, driving on freeways, living in large, comfortable homes. The next year we made a 10-day passage in our small boat in rough seas to small islands 1,100 miles from Australia, visiting people in tiny villages who subsisted on the food they grew in their gardens, and trading with visiting cruisers for their necessities. In between these trips we had a wonderful time cruising up and down the east coast of Australia visiting small towns and big cities, and enjoying all the advantages of life in an advanced country while living on a small boat. It was a lifestyle we enjoyed so much that we continued for eight years. We were happy to postpone the difficult decision of whether to continue sailing around the world or to sail back across the Pacific Ocean to Seattle.

## Cockroaches

The first time we saw a cockroach on board our boat was after our arrival in Australia in 1998, five years after leaving Seattle. We had cleared customs in October at Scarborough, just north of Brisbane. Over the next four months we worked our way down the coast to Port Stephens in New South Wales. We were planning to leave in another month to fly back to the States to visit friends and relatives.

A few weeks after we arrived in Port Stephens we sailed to the well-protected anchorage of Fame Cove. A low pressure system had formed off the coast, bringing strong winds and torrential rains for a couple of days. The second morning I woke up early to the sight of a large cockroach trying to climb into the cabinet just a foot above my head. I was able to kill it by spraying it with bug spray. We opened the cabinet and took everything out and found another cockroach crawling amongst our food stores at the other end of the cabinet. Now we realized we had a problem. How many more might be lurking in nooks and crannies throughout the boat? We put out a couple of roach hotels at the end of each cabinet, hoping to kill any remaining cockroaches. (Roach hotels are small black plastic domes filled with poisonous bait).

The winds had abated and the rain stopped, so we sailed back to

Salamander Bay that afternoon. The next day when we walked to the shopping center for groceries we bought six more roach hotels and three roach bombs. I taped the roach hotels down on the floor of every cabinet near the galley, behind the mast, and under the stove—everywhere we thought cockroaches might be roaming at night. For a week we didn't see any, and hoped our problem was solved. Then the next evening we saw a large cockroach near the galley, which I killed with bug, spray. Later, as I was stirring a pot of food on the stove for our dinner, I felt something crawling up my leg. Another large cockroach! Dan quickly squashed it against my calf. I took everything out of the food lockers and thoroughly cleaned the inside of the cabinets before replacing the food. I found another small cockroach under a potholder. That night I mixed up equal parts of boric acid and powdered sugar, placed a spoonful in each of four jar caps and placed these along the edges of the floor. We now had weapons all over the boat, but didn't know if any of them were working. We wanted to make sure that when we hauled *Shaula* out and flew back to the States we didn't have a colony of cockroaches multiplying inside our boat. We decided to set off a roach bomb—the only sure way of killing all the cockroaches.

A few days later we began preparing for the bombing. First we took all fresh fruit and vegetables, pots and pans, dishes, cups, glasses and silverware out of the cabinets, placed them in large plastic bags and put them outside in the cockpit. We covered all counters and cutting boards with newspapers, closed all the ports, opened up all the cabinet doors and hatches (33 altogether), and set the bomb in the center of the floor. I climbed out into the cockpit, and Dan set the bomb off, then quickly joined me and closed the main hatch. We left in the dinghy to go to the shopping center while the bomb was filling the boat with poisonous fumes. After two hours Dan rowed back to the boat and opened the main hatch and ports to air it out. For the next hour we read the paper and worked the crossword puzzle ashore. When we returned to our boat we found no odor, but did find eight small dead cockroaches on the floor. Over the next few days we searched the boat and found five more dead ones, but fortunately no live ones. The bomb apparently worked. Our boat was finally cockroach free.

The next time we saw a cockroach in our boat was 18 months later in Vanuatu. It was October and we had just made a grueling trip south against SE winds from Malekula to Port Vila. We had a problem with the stitching on our jib and had invited our friends Jim and Ann on *Insatiable* to come over for happy hour to help us figure out how to repair it. I was cleaning up the boat when I noticed a cockroach outside the vegetable crate next to the mast. I used insecticide spray to kill it, removed the crate to the cockpit and sprayed the floor under the crate. Next I lifted the lid of a tall Tongan basket tied to the mast near the vegetable crate. This was where the cockroaches had been nesting, hidden by the basket lid. Dozens of cockroaches swarmed frenetically at the bottom of the basket. I quickly sprayed the inside of the basket killing 20 to 30 of them but a few got away. We threw the basket full of dead cockroaches into the trash.

Over the next few nights we saw several more live cockroaches. It was time to set off another cockroach bomb. This time we knew how to do it. We set it off in the afternoon, and were invited over to *Insatiable* for happy hour while it sprayed. Two hours later we opened the boat and aired it out for an hour. Later that evening we killed eight slow-moving cockroaches that had been poisoned. A few days later we made a three-day passage to Noumea, New Caledonia. The saga wasn't over. Jim and Ann arrived in Noumea a day later, and we invited them over to happy hour to celebrate our passages. As we entered the boat after removing our shoes we spotted a large cockroach in the galley. I was horrified. Jim nonchalantly stomped on it with his barefoot to kill it (which I could never do). The next day we looked up the life cycle of cockroaches on the Internet and learned that we needed to set off another bomb two weeks after the first one to kill off the cockroaches that had hatched since. Once again we set off a bomb and left the boat for two hours. Dan came back to air it out, and we stayed on shore for another hour. After this cockroach infestation we never had a serious problem again. Occasionally we saw a large cockroach fly in through the main hatch, but we were able to kill it immediately.

It is almost inevitable that cockroaches will get onto a boat in the tropics. They come in on produce bought at the market, and fly in through the hatch. It is not shameful, and it is not due to slovenly housekeeping. In the tropics cockroaches are everywhere. In Brisbane we saw hundreds of them at night when we walked from the restrooms down the docks to our boat. They collect around the terrace at the marina restaurant, and crawl around the docks to feast on bait left by children who fish from the dock. We learned to do the cockroach stomp whenever we walked down the docks at night. It was easy to crush a dozen or more cockroaches on the way back to the boat from a shower in the evening. Roach hotels work to keep down the population on the boat, but for a real infestation the only effective treatment is an insect bomb.

## Leeches of New South Wales

A few months after our arrival in Australia in 1998 we sailed south down the Queensland and New South Wales Coast to Camden Haven. It was an easy overnight sail from our last port of call of Coffs Harbor. Departing at 4:30 pm, we stood three-hour watches, and came to the entrance of the river at 7:30 the next morning. The seas were smooth as we crossed into the river entrance, having timed our entrance for slack tide. We motored up the winding channel for an hour to the town of Laurieton, on the north shore of the river, nestled under a tall mountain called North Brother. We anchored in about 12 feet of water near the Returned and Services League (R.S.L.) club, had breakfast, and caught up on our sleep after the overnight passage.

Early the next morning we headed into town, planning to hike up to North Brother before it became too hot. It was December, the beginning of summer in Australia, and would be in the 80s in the afternoon. We walked to a service station to ask where the trail starts. It was just a few blocks up the hill from where we had left the dinghy at the town dock in front of the R.S.L. club. At the end of the road we reached a sign announcing Dooragan National Park, and a wide trail led behind a row of houses. From the level trail we took a branch off to the left that climbed up the hill towards the mountain, but after 20 minutes the trail ended at a fenced-off reservoir. Then we found a two-wheel jeep trail, which we took for another 20 minutes until it ended at an abandoned quarry. So much for climbing North Brother; we couldn't find the trail, so headed back to town for coffee, then back to the boat. That afternoon a young man came over in his dinghy to admire our boat, and we gave him a tour. He and his wife were teachers, and had a small boat called *Little Sister* at a nearby mooring. He invited us over for dinner that evening to their home at Bonney Hills. He met us in front of the R.S.L. club at 5 pm and drove us up the hill to the top of North Brother, where we delighted in the view of the valley below and the Camden River meandering down to the sea, with sand banks which could trap a boat if not carefully following the path between the buoys. We had a wonderful dinner and visit with Greg and Ann, and their daughter. Greg's parents joined us after dinner to talk about cruising across the Pacific.

Having seen the spectacular view from the top of North Brother we were determined to make the climb to the top on foot, so the next morning we again got up early, took our binoculars, and headed to the start of the trail. This time we knew we should stay on the straight level path for a half mile before it began to climb up the slope. It was a well-maintained trail, with only a few steep places with outcrops of rock. It took about two hours, and then we were on top and were rewarded with a great view from a viewing platform. It was well worth the climb.

While anchored in the river we enjoyed the other attractions of Laurieton—the R.S.L. club, where we were given a key for free showers in the restroom at the side of the building, several cozy coffee shops, a small library, and at the other end of town a grocery store and a wonderful Thai restaurant. The level trail at the base of North Brother was one of our favorite places to stroll along under the trees to look at birds—parrots, lorikeets, and wompoo pigeons. Then we had three days of rain, turning the trail somewhat muddy and not inviting. A week after our first climb the rain stopped. It was ANZAC day, which is celebrated by a ceremony on top of the mountain. We decided to climb North Brother again, this time with Ted and Sue, who had arrived several days earlier. They were Australians who had emigrated from England and now lived on a 30-foot sailboat, the *Alice Colleen*. As we started up the trail we began to find small slug-like creatures climbing on our shoes. We were told they were leeches. They were about a half-inch long, ugly, and leathery. Apparently, as our feet hit the ground they were stimulated by the vibrations to latch onto our shoes, then climb onto our socks and legs, seeking warm blood. They were not easily visible, but if we bent down to look closely we could see an occasional thin black body sticking up from the ground and waving its head back and forth. Every 10 or 20 minutes we checked, and found one or more attached to the side of our shoes or socks with their suckers. They could be removed by prying them off with a stick, before they reached the skin to begin sucking blood. When we reached the top of the mountain two hours later all of us found more leeches on our shoes and socks. We sat down at the picnic table and used sticks to remove them. We were horrified that we could be attacked by leeches when we were innocently hiking through the forest to look at birds and trees. Unfortunately, we had missed the ANZAC ceremony, which began at dawn.

## Haulout in Oyster Cove

Every two years during our cruise across the Pacific we planned to fly home to visit Dan's parents in California, visit friends and relatives in Seattle and take care of our doctor, dentist, and dermatologist visits. It was important to find a safe place to leave the boat while we traveled home, preferably hauled out on solid ground, rather than in a berth in a marina. When our boat was hauled out we didn't have to worry that the boat would sink while we were gone.

This was our first haulout in Australia. A cruising couple we had met a few months earlier told us they had hauled their boat out at Oyster Cove several times, and thought it was a well-run family operation. It had a secure fenced-in area for long-term storage. About a month before we were to leave we called up the yard to make a reservation for haulout and storage. The yard is located on a narrow inlet on the far west side of Port Stephens, entered through a shallow channel. It can only be approached at high tide when the depth of the water is over seven feet, so the timing is critical. They gave us a date of March 17 in the morning for the haulout. We made our plane reservations for the afternoon of the 17$^{th}$ to fly from Sydney to Los Angeles.

Instead of a travel lift for hauling the boat out they had an ingenious tractor-trailer system, which Mr. Bailey designed and built himself. A travel lift is a huge box-like steel frame on wheels. Its powerful motor lifts boats out of the water by means of wide straps. The straps are positioned under the boat and they lift the boat high enough to clear the ground. The operator drives the travel lift with the boat suspended in the slings into the yard where it will be stored. It is always a heart-stopping moment watching boats suspended high above the water on two slings. One worries that the straps are in the wrong place, that the boat will be lopsided in the slings, that a strap could break, or the rigging could be scraped and damaged on the top rail of the travel lift. The tractor-trailer system seems safer, because the boat sits solidly in a cradle on a trailer and is gently towed up out of the water on a ramp instead of being raised 10 feet above the water.

To be sure we would make it in time for the haulout we arrived at Port Stephens 10 days before the scheduled time. We anchored at Salamander Bay, where we could buy groceries and visit an Internet café. Two days before our haulout we had to motor our boat up the inlet at first light, when the tide would be high enough that we wouldn't go aground. The channel was well marked with red and green buoys. It took only 45 minutes to reach Oyster Cove, where we anchored and went ashore to meet the Baileys, and look over the haulout yard. The first thing we noticed was that the secure fenced-in storage area had very few boats. There were a dozen or more boats hauled out in the yard where boaters work on their boats. Mr. Bailey told us that a couple of weeks before a gang of men had climbed over the fence of the storage area and broken into several powerboats, stealing their fish-finding equipment and other gear. None of the sailboats had been broken into because they are higher off the ground, and there was no ladder. The Baileys had moved most of the boats out of the secure area and into the yard, which was not fenced in, but was closer to their house. They thought they would hear if anyone broke into the boats.

While at Oyster Cove we used up our fresh food, faxed the necessary forms to customs for leaving our boat in Australia, and packed our bags for the flight home. The morning of our haulout we motored over and tied our boat to the dock adjacent to the ramp. The trailer, which had been used to haul out another boat earlier that morning, was parked at the top of the ramp. After fixing some sandwiches we walked up the ramp to eat our lunch on the grassy area next to the parked tractor. Suddenly we heard a loud explosion, and water began gushing out of the tire on the rear of the tractor. Luckily it was on the side away from where we were sitting. A large hole was blown out of the tire, and the water that is used to stiffen the tires of the tractor was all leaking out. The bolts holding the tongue of the trailer to the tractor broke off when the tire blew. Mr. Bailey was unfazed. He simply took the tire off, drove to the nearest small town, and came back an hour later with a repaired tire. He installed the tire and replaced the broken bolts on the trailer tongue. Our faith that this was a well-run operation, however, was eroding. Our haulout a half hour later went smoothly. Mr. Bailey's son drove the tractor over to the area where our boat would sit for the next four months, set it down, and propped it up with two jack stands on each side, and one in front under the cut-away forefoot. I noticed that the jack stands on either side of the boat were not chained together, as is the standard practice. This prevents the jack stands from sliding out away from the boat, and the boat tipping over. When I mentioned this to Mr. Bailey, he said it wasn't really necessary.

Our van arrived shortly afterwards, and as we left I noticed with great relief that Mr. Bailey was connecting the chains underneath the boat to the two sets of jack stands on either side. (He had been kidding when he said it wasn't necessary.) On our two-hour van trip to the Sydney airport we were the only passengers. We had time to think about our situation. We wondered if we had made the right choice of a haulout area. Our most valuable possession, our boat, which had carried us thousands of miles across the ocean and been our home for the last six years, would be on the other side of the world while we visited the U.S. for four months. Would a thief break into the boat in the middle of the night and steal our electronic equipment while the Bailey family slept? Would a huge storm knock the boat over onto the ground? Would a few cockroaches survive the cockroach bomb and multiply into a large colony while we were gone? It would be ironic that after surviving gales and rough seas as we crossed the ocean our boat was damaged when we left it on land. It seemed best not to contemplate all the disasters that could happen, and just enjoy the scenery and think of the nice long visit with friends and relatives in the States, trusting that our boat was in good hands in Australia.

## Grounding in Pancake Creek

After returning from our visit to the States we ventured up the Queensland coast to visit the Whitsunday Islands in August 1999. The Whitsundays are a group of beautiful tropical islands about 700 miles north of Brisbane. These islands, with clear warm water, white sandy beaches, and many protected anchorages, are favorite cruising grounds for Australian and foreign sailors. On the trip north there are good anchorages all along the coast. Most trips are easy dayhops, the anchorages being less than 50 miles apart.

At Bundaberg, just 200 miles north of Brisbane, we stayed in a marina for several days, and shared dinners and happy hours with cruisers on several boats—Carl and Sandy on the Valiant 40 *Fantasea*, Jack and Wanda on *Sequester* and Jim, Janice and their daughter Rhiannon on the catamaran *Pas de Chat*. All of us were planning to sail to the Whitsundays over the next few weeks. From Bundaberg to Pancake Creek was one of the longest passages, about 60 miles. We could make the trip during daylight hours if we left early in the morning. All three sailboats were planning to leave Bundaberg the same day for Pancake Creek. It would be nice to sail in company.

At 6 am, before sunrise, we left the marina, motored out the Burnett River and started up the coast. At first there was not enough wind to sail, so we had to motor. In the afternoon the wind picked up and we had an enjoyable sail downwind with the full mainsail and the jib poled out on opposite sides—wing and wing. We were making good progress, averaging over six knots, which meant we would reach the anchorage well before sunset. *Sequester* and *Pas de Chat*, who were sailing along the coast with us, decided to peel off to anchor at Round Head in mid-afternoon. The entrance to the Round Head anchorage was over a shallow bar, and with our five-foot draft we wouldn't be able to cross the bar until close to high tide, around 8 pm. We decided to continue on to Pancake Creek as planned. *Fantasea*, a larger, faster boat continued on past Pancake Creek towards an anchorage further north. Now Dan and I were all alone. However, with the wonderful following wind we were enjoying the sail, and looking forward to the well-protected anchorage at Pancake Creek.

Finally, at around 5 pm, we reached the entrance of Pancake Creek. The channel into the anchorage is S-shaped. We entered at the top of the S and curved around to the left in a deep channel well marked by buoys. The channel narrows in the middle of the S. The straight path through this narrow channel is marked by two range markers, one on land, and the other further out in the water. To stay in the middle of the channel we had to keep the two range markers lined up. This was easy. However, our guidebook showed that somewhere along this range you have to turn to the right and curve around the bottom part of the S to reach the anchorage. On a sunny day, with the sun shining through the water it would be easy to know where to turn, as it would be clear enough to see the rocky or sandy shallows on either side of the channel. Now with the sun low in the sky, the water looked the same everywhere. We were on our own, with no idea where to turn.

We made a guess, turned to the right and immediately struck a rock. The boat came to a sudden halt. The crunch of our hull on the rock sent a jolt through my body. Had we turned too soon, or too late? There was no way to tell. After a few minutes the tide rose enough to float our boat off the rock. We lurched forward a few feet and struck another rock. After a few minutes the tide rose enough to again float us off, and we struck a third time! I was panicked. I had visions of remaining on the rocks all night, the boat heeling over as the tide went out. I hate going aground. To me it was like driving down a road and running into a ditch, this time a ditch strewn with boulders. However, we could console ourselves that it wasn't our fault, but a lack of buoys to lead us into the anchorage that caused us to go aground.

The only way we could find the channel was to climb into the dinghy and poke an oar down to the bottom to determine how deep it was on either side of our bow. We quickly launched the dinghy from the cabin top, and I rowed out about 10 feet forward and to the left of where we were aground, and sounded the bottom with the oar—it was less than three feet deep. When I rowed out to the right and sounded with the oar it sank to length of the oar—so the channel was on our right, and we had turned too late and hit rocks near the shore. A few minutes later the tide lifted our boat off the bottom for the third time, and we veered off to the right and into the deep channel with great relief. From there it was only a short 10-minute trip through the curving channel to the anchorage. Our grounding had caused a short delay and much anxiety, but we were happy we were finally able to drop our anchor in the calm, secure anchorage. It was just before sunset.

Going aground on sand is bad enough, but going on rocks can be catastrophic. Not only is it difficult to get off, it could do serious damage to the bottom of the boat, causing gouges or cracks in the fiberglass. We were glad we had made it into Pancake Creek before high tide. The rising tide made it possible to float off the rocks repeatedly. There was no way of knowing what damage hitting the rocks had done to the bottom of the boat. We would assess the damage the next time we hauled the boat out, some time next year. Meanwhile, we would continue another 500 miles up to the Whitsundays, trying to forget about our ill-fated entry into Pancake Creek.

## Lamington National Park

Two years after our arrival in Australia we needed a break from the boat. Our boat was in the marina in Brisbane, where we had spent a month sanding and repainting the bulwarks and bowsprit, cleaning the topsides and polishing the stainless steel stanchions. We'd had our dodger re-stitched, the jib re-stitched, the chain and anchor re-galvanized, and had sent our laptop computer back to the States for repair. It was December, the start of summer (the equivalent of June in the northern hemisphere) and Brisbane's weather was becoming unbearably muggy. Some cruising friends recommended a visit to Lamington National Park in Queensland. A trip to the cool rainforest in the mountains to hike and look at birds seemed perfect. We invited our friends Gordon and Miriam to go with us. They had only arrived in Australia a month before, but like us had fallen in love with the birdlife. Quite often we took walks around the neighborhood of the marina in Brisbane, and saw a wonderful collection of colorful birds. Many mornings at the marina we were awakened to the call of butcherbirds that landed on top of the masts and serenaded us with their melodious songs. Lamington Park is home to a huge variety of birds, and we were excited at the prospect of visiting our first rainforest.

It was an easy two-hour trip by train from Brisbane to Nerang, where a courtesy bus from the Binnaburra resort picked us up to take us up to the mountains. We had reserved two tent cabins in the campground in Binnaburra for three days in mid-December. The tent cabins were quite luxurious—a large wood-framed cabin with canvas walls, a queen-sized bed, table and chairs, and a terrace in the front with a great view across the valley. The meals at the nearby Lamington Tea House were delicious and very reasonably priced. It was less than half the cost of staying at the nearby Binnaburra Lodge or at the more upscale O'Reilly's Guesthouse on the other side of the park.

Early the next morning we were awakened to an orchestra of birdcalls, many of which we had never heard before. There were chirps, tweets, melodious tones, squawks and a repeated "mew mew" sounding like cats. Dan and I took an early morning walk on a two-kilometer rainforest trail before breakfast. The first thing we saw were numerous small wallabies, called pademelons, hopping around and grazing on the vegetation near the tent cabins in the campground. They were the smallest wallabies we had ever seen. They quickly disappeared after the sun came up. We had breakfast with Gordon and Miriam at the Lamington Tea House and planned a hike on Dave's circuit trail for the rest of the day. Armed with binoculars, cameras, and knapsacks we strolled down the trail observing a large variety of birds in the trees and on the ground. It was necessary to look down every few minutes to avoid tripping on the many tree roots crisscrossing the trail. We were delighted when Miriam spotted one of the elusive birds that we had heard mewing in the tall trees. We identified it as a green catbird, and watched it through our binoculars as it repeatedly mewed like a cat. We saw the whipbird and heard its call like a whip whooshing through the air followed by a loud crack at the end. We saw many rufous fantails flitting around and spreading their tails like a fan when they landed, blue-black satin bowerbirds with beautiful blue eyes, large red and green king parrots, crimson rosellas, bright red with blue wings and tails, laughing kookaburras, and small bush turkeys running through the brush. It was a birdwatcher's paradise.

Halfway through the hike Miriam's boots began to disintegrate. Soon the sole at the toe end separated from the rest of the leather, resulting in flapping of the soles with every step. Dan attempted to solve the problem by tying the soles firmly to the top of the boots with string that he always carries with him. Unfortunately, shortly afterwards, Miriam tripped and fell flat on her face, the camera around her neck pressing into her chest. It was so painful she thought she had broken a rib. She carried on bravely for the rest of the hike, but with pain at every step. We made it back okay and had another delicious meal at the Lamington Tea House, followed by wine out on the terrace of Gordon and Miriam's tent cabin while we planned our next day's hike.

Again we woke early in the morning and all four of us walked over to the vicinity of the lodge to look for birds. We ran into a small group of birdwatchers from the lodge led by an ornithologist. He welcomed us to join in, and we spent a half hour seeing many more new species to add to our list of Australian birds. After breakfast the four of us set out on a six-kilometer hike on the Tangawalla trail. Miriam had to abandon her boots and wear tennis shoes for the rest of the time, and was in a lot of pain from her fall, but she soldiered on. Again we saw some amazing birds such as golden whistlers, rainbow lorikeets, multi-colored noisy pitas, and the brown cuckoo dove with its iridescent neck and long tail. Near the end of the trail we were treated to a large circle of Antarctic beech trees that were 200 years old, growing as suckers from 5,000-year-old roots of long-gone ancestors. It was truly an ancient forest. We finished the day with another delicious dinner at the teahouse.

The third morning we took another bird walk around the campground area, then Dan and I took a two-hour walk on the Caves track through a dense rainforest with many huge trees, but saw only a few birds—firetails and scrub birds. That afternoon we had to pack up and catch the bus and train back to the marina.

It was a wonderful vacation from living and working on a small boat in the heat of a Brisbane summer. Three days of hiking, seeing an amazing variety of birds, and eating delicious meals without having to cook or wash dishes on our boat was heavenly. Miriam had X-rays when we returned to Brisbane, and found she had no broken ribs, just a very painful bruise. We vowed to make another visit as soon as we could to this delightful rainforest.

## Queensland Fauna

We sailed south from Brisbane in the last week of January 2001, planning to stop at Southport to visit a couple of nearby wildlife parks—Currumbin Wildlife Sanctuary and Fleay's Fauna. Along the way we anchored at Tangalooma, famous for its dolphins. Every afternoon at 4 pm visitors and people staying at the resort begin lining up on the beach in four or five long lines about 10 feet apart to feed the dolphins. A pod of dolphins that have learned they have a guaranteed source of fresh fish every afternoon comes in near the beach at the scheduled time. We watched as the people nearest the water in each line were given a fish from a bucket to hold out to a dolphin, which took it out of the person's hand. Then the next person in line was given a fish to feed a dolphin. This continued until all the fish were gone. It was amusing to watch how thrilled people were to feed the dolphins. Near the resort there is an excellent marine research center that we visited the next day to learn more about dolphins and watch videos of manatees and dugongs.

After Tangalooma we sailed south to Southport. It was the last day of the school holidays, and the anchorage was very crowded. It was a crazy scene, with powerboats and jet skis, taking joy rides in the small bay, roaring through the anchorage, causing lots of noise and wake. We decided to move to a quieter place. The next day we went ashore in our dinghy to the nearby beach, and walked down along side the highway to the Southport Yacht Club. The yacht club house is surrounded with grass, and we watched some wallabies grazing on the lawn. They kept the lawn well-trimmed, with no need for a lawn mower. We walked further down the road to an open-air bar, set back from the beach, surrounded by trees and many wallabies. The owner let us use his phone so we could call the yacht club and reserve a berth for a week. While we had a glass of wine we were delighted to watch the resort owner feed a pair of kookaburras hotdogs cut into one-inch lengths. The birds were like pets that came every day to get their free handout.

The next day we motored down to the yacht club and tied up at our assigned berth. From the yacht club it is only a short walk to the Marina Mirage shopping center. We walked down nearly every morning to have coffee and read the *Sydney Morning Herald*. The main attraction of the shopping center was the large aviary two stories high. It was filled with rainforest birds, many of which we recognized from visiting Lamington National Park. It was a microcosm of a rainforest in which we could observe the many colorful birds flying from branch to branch of tall trees from the second floor of the mall, and ground dwelling birds on the lower level, and hear their varied calls. We didn't do much shopping, but we always stopped to observe birds in this free aviary.

From the Southport Yacht Club marina we took an hour bus ride to Currumbin Wildlife Sanctuary, leaving in early morning. Currumbin started out as a floral garden and honey farm, but it soon attracted rainbow lorikeets, which in turn attracted visitors. We were given small dishes of honey water that we held out to the birds. Dozens of rainbow lorikeets flew down from the trees to land on our arms or hands to slurp up the honey solution. In another part of the sanctuary we watched the feeding of dozens of kangaroos and wallabies gathered at long troughs on the ground. It was the largest collection of kangaroos and wallabies we had ever seen. Another attraction was the koala exhibit, where we could take photos of a koala held by a zookeeper. Koalas look cuddly, but they have such sharp claws that the attendant has to wear a heavy vest to avoid serious scratches. We watched the feeding of Tasmanian devils, which are rare animals resembling small black pigs. To make their lives more interesting the zookeeper hides their food, which they find by smell. In addition, we visited several small aviaries to see the many colorful tropical birds.

Our next trip was by bus to Fleay's Fauna. This park has miles of trails leading to all the various attractions. We went to the large aviary, filled with tropical plants and beautiful parrots. As soon as we entered the double door to the aviary we noticed a satin bowerbird flying towards our faces. It appeared to be homing in on Dan's blue eyes, but then flew past us to the young woman who came in behind us. She had her blue-tinted sunglasses propped on the top of her head. With a quick scoop the bird snatched the sunglasses off her head and flew to the far corner of the aviary where he was constructing his bower. It happened so fast she couldn't believe it. She was understandably upset—she said they were very expensive sunglasses, and she couldn't afford to replace them. Dan offered to retrieve them. He took the narrow path along the side of the aviary towards the bower, but as he approached the bower where he could see the sunglasses, the bowerbird attacked him repeatedly, and drove him back. He apologized to the lady, and suggested she ask the ranger to help. Among my favorite animals was a tree kangaroo sitting in the notch of a tall tree like a lanky brown cat. There were wild brolga cranes and jabirus, and barking owls. We attended the crocodile feeding and the pelican feeding. We watched a platypus dive down in the large aquarium and fill its cheek pouches with earthworms as it floated to the surface. Fleay's Fauna was the first place to breed platypuses in captivity.

In the afternoon visitors were allowed to feed the birds in the aviaries. We returned to the aviary we had visited that morning with our bags of food to feed the colorful eclectus parrots—the females bright red with blue wings, the males bright green. The male and female plumage is so different that for many years they were thought to be two different species. As we were feeding the birds another young woman walked in the door with sunglasses pushed on top of her head, and the same thing happened as that morning. The satin bowerbird quickly snatched them off and flew to his bower. This time the park ranger was there and knew how to outsmart the bowerbird. She carried a pocketful of bright blue buttons. As she approached the bower she tossed a button onto the ground to distract the bird, and grabbed the sunglasses before the bird could get back to the bower. This obviously was a frequent occurrence. We had seen a bower of the satin bowerbird in a park in New South Wales with an amazing array of blue objects—plastic bottle caps, blue drinking straws, and blue plastic clothes pins. They are attracted to anything that is blue, and will even steal these objects from another bird's bower.

As we cruised along the coast of Australia we avoided most tourist attractions. Besides the friendly people and the wonderful coastal cruising, the exotic birds and animals were a huge attraction for us. It was like our Australian friends who visit America and are thrilled to see bears in the mountains, or squirrels in the city. We couldn't resist the displays of Australia's unique birds, mammals, and reptiles. We were delighted to spend a whole day at each of these wildlife sanctuaries.

## An Extra-tropical Cyclone

Every two years during our South Pacific cruise we planned to fly back to the States for a four- or five-month visit with Dan's parents in Southern California and friends and relatives in Seattle. In February 2001, about a month before our second visit to the States from Australia, we sailed down to Yamba in New South Wales. Yamba is a small town near the mouth of the Clarence River, fairly close to the border with Queensland. It has everything we wanted—a small marina, a haulout yard and storage facility, friendly people, an Internet café, good restaurants, and shops for groceries, seafood, and produce. Our plane reservations were for March 14$^{th}$ from Brisbane to Los Angeles. We planned to have the boat hauled out and put in storage on March 12$^{th}$, stay in a nearby motel overnight, and take the six-hour bus ride to Brisbane the next day. Meanwhile, we took turns having dinner on our boats with friends Jim and Lynn or shared meals at Thai, Indian, or seafood restaurants, saw movies at the local movie theater, and swam at the nearby beaches. We met Jim and Lynn the previous year just before departing New Caledonia for Scarborough, Australia. We made the crossing to Australia at the same time, and since then enjoyed getting together as we traveled down the coast to Brisbane and Yamba. We especially enjoyed taking bird walks to see the many colorful birds around Yamba such as rosellas, cockatoos, rainbow lorikeets, honeyeaters, wattlebirds, and dollar birds. After two weeks Jim and Lynn had their boat hauled out and left for Olympia, Washington.

A week later it began to rain heavily and blow 30 knots or more. This continued for three days. Then a deep low that had formed off the coast moved inland and passed right over Yamba. It brought torrential rain and winds gusting to 70 knots. This was an extra-tropical cyclone. All night the wind shrieked in the rigging, rain pelted against the side of the boat, and gusts heeled the boat over repeatedly in the slip. In the higher gusts the boat was knocked over about 45 degrees. To determine the wind speed we had to open the hatch, and one of us stick our head out into the cockpit to read the anemometer. Each time we were blasted by wind and rain. It was impossible to sleep. This was the strongest wind and heaviest rain we had ever had on the boat. We always avoided being in the tropics during cyclone season. The maximum winds we had experienced while sailing was 40 knots, with 15-foot seas. It was frightening to imagine being caught out at sea in a cyclone with 70-knot winds and 30-to-40-foot waves.

The next day we learned it had rained 12 inches in Yamba in the previous 24 hours. The Clarence River was rising, soon to overflow its banks. When we took walks between rain downpours we could see the flooding of the yards of homes across from the marina. The river was roaring past Yamba down to the sea, and had turned coffee-colored. Trees, branches and other debris were floating down to sea. We walked out to the beaches where we had swum in clear calm waters a week before. Now the beaches were deserted, with no swimmers or surfers. Chocolate-colored waves were breaking on the ocean beach. After several days of the muddy river water running out to sea, the ocean turned a brown color as far as the eye could see.

Yamba was completely isolated from the rest of the world. The only highway out ran along the south shore of the Clarence River, and it was completely submerged. This was the worst flood in 100 years.

However, the local residents experienced some flooding every few years, and knew what to do. Every morning men from Emergency Services maneuvered a large powerboat 40 miles down the river from Grafton bringing bread and other supplies to the main dock at the Yamba Marina. Kevin, the marina owner, helped unload hundreds of loaves of bread from the powerboat to the dock, to be taken by truck to the shops in town. People who had medical emergencies were picked up by helicopters from flooded towns along the river and flown to the Grafton hospital.

When we learned that the main highway was underwater we realized that we would be unable to get to Brisbane to catch our plane. I called the Flight Center where we had purchased our tickets to try to change our tickets to a later date, but the saleswoman told me there would be a $300 fee for canceling and reissuing new tickets. She said to wait until it was closer to the departure date. Hoping the flood would subside soon we went ahead and had the boat hauled out and put in storage as scheduled. We took off the luggage we needed for our trip, put everything on deck in the cabin, closed up the boat and walked across the street to check into the nearby motel. Unfortunately, the roads were not expected to be passable for days. Again I called the Flight Center to book a later flight, but she wouldn't change the date without charging us a fee. We felt we shouldn't have to pay a fee for conditions that were not our fault. Finally Kevin, the marina owner, suggested we talk to the local constable, who might be able to help. We walked down the street to the police station and told the head constable our plight. He got on the phone with the girl at the Flight Center and said authoritatively "This is Constable Hutton. I have an American couple here who are stranded in Yamba. It's impossible for them to get to the Brisbane airport for their flight due to a flood." She finally yielded, and allowed us to change our ticket date for free. We picked an arbitrary date five days later, hoping that the waters would recede by that time.

Fortunately, three days later, the highway was finally opened. We caught a bus to Grafton, and there transferred to another bus to Brisbane. On the bus trip we were amazed to see flooded cane fields all along the south bank of the river, hundreds of trees knocked down, and houses surrounded by a sea of water. The roads were still under a foot of water in places, but the bus was able to get through, and we made it to Brisbane for our flight.

Although cyclone-force winds and river flooding happen regularly in Australia, we had never experienced either of these catastrophes. It was a devastating cyclone, but fortunately no one was injured or killed. Although we were frustrated that we were unable to leave when we had planned, in the end a delay of six days didn't matter. We were just happy to have survived a cyclone and a flood unscathed. Fortunately our boat was safely tied up in a marina instead of being at anchor or out at sea.

## Visit to the U.S. 2001

When we left Yamba in mid-March 2001 we planned to stay for six months in the U.S. For two nights before departing we stayed in a youth hostel in Brisbane, which gave us a chance to visit friends on their boats in the marina in Brisbane. Taking an airporter to the Brisbane airport, arriving two hours early for the flight, having a 30-minute layover in Auckland, a long wait to clear customs and immigration in Los Angeles, and taking another airporter from Los Angeles to Glendale added about eight hours to the 14-hour flight time. We arrived in Glendale at 11:30 am (5:30 am Australian time) completely exhausted, having been on the go since 7:30 am with very little sleep. It was not easy to travel this way in our mid-60s, which explains why we only traveled to the U.S. from Australia every two years.

Dan's folks were glad to see us, and we had a great visit. At 89 years of age, they were still living in their house, although finding it somewhat hard to manage without a lot of help. For five months we helped out with household and gardening chores, and helped move all the furniture outside for a day so the carpet could be replaced. In July we took a train up to Seattle for a month's stay, where we visited friends and took care of our medical and dental checkups. On the train trip back to Glendale we stopped in Eugene, Oregon to celebrate our nephew's wedding. My sister and brother-in-law came from Alabama, and my other sister came down from Seattle to celebrate. It had been over 10 years since we had all been together, and we had a wonderful reunion. Reminiscing about our childhoods in Knoxville we decided we should write a book, called *The Three Middle Sisters*.

After returning to Glendale we had six weeks to visit and take care of Dan's folks' needs—taking them to their medical appointments, and getting them set up with Lifeline so they could get help if they fell at home. We departed at 7 pm for the airport for our flight back to Brisbane on September 10$^{th}$. The flight was smooth, and we landed in Auckland September 12 at 5 am New Zealand time (10 am California time). We had a short layover before boarding the flight to Brisbane. We wandered down the main corridor of the airport, where we saw a large crowd standing around a TV screen next to the board listing arrival and departure times for all flights. People were gawking at the TV with stricken faces, uttering exclamations of disbelief and horror. On the screen were scenes of a plane crashing into the Twin Towers, flames escaping out of the upper floor windows, and people plunging to their death. We were stunned. People were saying: 'I can't believe this is happening. It's like a movie or a video. It can't be real!" The terrorist attack had occurred just a few hours before while we were flying across the Pacific Ocean. We continued to watch transfixed by the horrible scenes on the TV screen until our flight to Brisbane was announced, then returned to our plane for the final leg of our journey. It was a somber day. The images of the burning towers and people jumping to their deaths kept re-playing in my mind.

We arrived in Brisbane later that morning and in MacLean in the afternoon after a six-hour bus ride south from Brisbane. We were still in a state of shock. The owner of the motel was waiting for us at the bus stop at MacLean to take us the 10 miles to the motel in Yamba. He expressed his sympathy for our suffering through this tragedy. We were touched by how supportive the Australians were. We crashed at the motel, recovering from the long trip back to Australia.

The next morning the marina manager took our boat out of the storage area and moved it to the yard where we could begin to work on it. The next two weeks we were extremely busy, working on the boat mornings and afternoons, returning to the motel in the evenings to cook dinner, have showers, and watch TV. Keeping busy during the day kept our minds off the terrorist attack, but in the evenings we watched CNN, where the whole catastrophe was repeated over and over. We watched specials on Osama Bin Laden, a special on the terrorist attack, and memorial services for the victims. The stock market fell 600 points to 8900 in one day, the largest point loss in history. All the news was depressing.

There was more work to do than our usual annual maintenance. Our boat had developed blisters on the hull which we had first noticed four years before when we hauled out in New Zealand. Dan had to drill out the gel coat around the numerous weeping blisters, fill them with epoxy filler and paint them over with waterproof paint. I spent days cleaning the hull above the waterline with stain remover, waxing and polishing the topsides, then polishing all the stainless steel fittings on deck. Dan had to remove the rudder, and have the mechanic in the yard help him replace the cutlass bearing and the stuffing box. Before re-launching the boat we carried out the usual tasks of sanding the hull, painting on two coats of anti-fouling, replacing the zincs, and greasing the seacocks. The boat was in great shape when we launched her after two weeks of intense work.

A few days after our arrival the young woman who owned the Internet cafe told us we could use the Internet as much as we wanted for free to contact friends and relatives in the States. She said it was the one thing she could do to help. We were grateful. We could imagine how frightened people in the states were over the possibility that there would be another attack. We felt relatively safe on the other side of the planet. Australians didn't seem to have enemies who were willing to kill themselves for their cause.

It was fortuitous that we had scheduled our flight from Los Angeles on September 10. After September 11$^{th}$ all flights out of Los Angeles were cancelled, and we would have probably been unable to leave for a week or more. Once we were back on our boat life became more normal for us, no longer having access to television. Our friends Jim and Lynn on *Windchime* had returned from Olympia and were working on their boat in the marina, and we continued to do some jobs on our boat in the water. We shared dinners on our boats and ashore, and the four of us took some wonderful birdwatching walks in the woods around Yamba. The horror of the Twin Towers attack gradually faded into the distance, and we could think and talk of other matters.

## Coffs Harbor

After our haulout in Yamba we sailed further down the New South Wales coast to Coffs Harbor, a small coastal town that's very popular with tourists. Coffs Harbor has a beautiful beach, many restaurants, and a shopping center that we reached by walking along a creek adjacent to a large botanical garden. It has a small well-protected marina from which we had departed when we sailed to Vanuatu the previous year. At that time we were so busy getting the boat ready for the offshore passage that we saw very little of the area surrounding Coffs Harbor. This time it was a stopping off place on our way south to Sydney, and we had a month or so to explore the area. When we first arrived Dan and I took an early morning walk through the extensive botanic gardens, where we saw a wonderful variety of birds—variegated fairy wrens, kookaburras, bowerbirds, sacred kingfishers, blackfaced monarchs, and orioles. Then suddenly in the middle of the forest we spotted three koalas—a very rare sight. A large male koala at the base of a eucalyptus was making a loud grunting sound, which we presumed was a mating call. We saw another adult koala clambering up a nearby tree, and watched a baby koala climb to the top of a very tall eucalyptus. This was a special treat to see koalas in the wild. Until then we had only seen them in zoos.

Jim and Lynn, on the boat *Windchime*, arrived at Coffs Harbor a few days after our arrival and, like us, wanted to do some sightseeing. We signed up for a half-day van tour around Coffs Harbor. Our driver was a fund of information. He drove us up to the plateau above the town with a panoramic view of the harbor and the ocean, and to a huge strangler fig tree, better known as a banyan tree. These trees start from a seed that has been carried by a bird to a nook in a tree, where it sprouts and sends out roots down the trunk of the tree into the ground. They are fast growing, and soon they completely surround the host tree with their roots, trunks and branches. After many years the original tree dies, and the strangler fig remains as a huge multi-trunk tree, hollow on the inside. The bus tour took us through miles and miles of banana plantations. Bananas were planted by the German immigrant who founded Coffs Harbor in the early 1800s. Although bananas were growing well in tropical Queensland, no one had tried to grow bananas this far south in the temperate climate of New South Wales. They were very successful. Now groves of banana trees cover the slopes all along the highway. According to our driver, most of the banana plantations are owned by families originally from India. The last part of our tour was to the large Sikh community in Woolgoolga, just north of Coffs Harbor. We were welcomed into a large Sikh temple where we were given a tour of the premises. Men with turbans and women in colorful saris were having lunch at a communal table in the main part of the temple. We were invited to share their lunch. The food looked wonderful, but we had to decline because our tour was over and our van driver wanted to get back to Coffs Harbor.

The following week was Thanksgiving, and we, Jim and Lynn, and new friends Austin and Patricia decided to celebrate by having dinner at the Bush Turkey Restaurant just a short walk from the marina. With a name like that it seemed appropriate. However, instead of turkey Lynn ordered kangaroo steaks, I ordered emu medallions, and the rest of our party ordered fish. The emu, a large Australian flightless bird, tasted like chicken, the kangaroo tasted like venison, and both were delicious. It was far from a traditional Thanksgiving dinner, but we enjoyed the new taste sensations.

Our next tourist adventure was a visit to Dorrigo National Park, a rainforest park just 20 miles inland from Coffs Harbor. Jim, Lynn, Dan and I rented a car to drive to the park. None of us wanted to drive because of the difficulty of driving on the left side of the road, but Dan had had the experience of driving a rental car for a month on the South Island of New Zealand, so he reluctantly agreed. After a half hour of driving along the narrow highway Dan noticed that cars were piling up behind us because of our slow speed, so he pulled over onto the left shoulder to let them pass. We came to a sudden stop with a loud bang. Dan had hit a low curb with the left front tire and blown it out. Luckily there was a spare in the rear. Dan and Jim replaced the tire so we could continue on our way. Dan was greatly relieved when Jim offered to drive the rest of the way. All was fine until we reached the last road leading to the park. Jim signaled a left turn with the lever on the left side of the steering wheel, which turned on the windshield wipers. (Dan had done this numerous times in our rental car in New Zealand). We were laughing so hard that we didn't notice Jim had turned into the right lane, which is easy to do when there is no traffic to follow. Now we were all yelling: "Left lane! Left lane!" Jim quickly changed lanes. Fortunately, there was no traffic headed our way.

Once at Dorrigo we took a short walk on the skywalk, above the forest canopy, then a long hike on a trail through the rainforest. The rainforest was filled with birds, a few that we had never seen before, notably the topknot pigeon, a logrunner and a Bassian thrush. We saw many bush turkey mounds—three-foot high piles of leaves, twigs and dirt that the male turkey builds to incubate the eggs. He mates with many hens, which all lay their eggs in his mound, and he continuously tends the nest by kicking the composting debris around to keep the eggs at just the right depth so the temperature remains ideal. This is very efficient, and convenient for the females, who are freed of the duty of sitting on nests to hatch their eggs. Further down the trail we suddenly stopped in our tracks when we heard the remarkable call of a lyrebird—the best imitators of all the birds. This lyrebird was singing continuously with one call after another, imitating all the birds in the vicinity one by one. Lyrebirds sometimes incorporate man-made sounds, such as chainsaws or outboard motors starting up. The lyrebird continued to sing for more than a half hour, never seeming to repeat itself. We had seen lyrebirds before with their beautiful long forked tail curved at the ends to resemble a lyre, but this was the first time we were treated to its remarkable song. It was the highlight of our birdwatching experiences.

Most of the time when we were in Australia we sailed along the coast seeing miles and miles of beaches and forests of eucalyptus trees or stayed at anchorages or marinas and visited towns or cities along the shore. We learned there is so much more to see if we were willing to travel a short distance inland. We were glad we had the time to see some of the inland parts around Coffs Harbor, even though it meant being tourists instead of cruising sailors.

## Camden Haven

One of our favorite anchorages on the east coast of Australia is Camden Haven. Like many of the good anchorages in New South Wales it is miles up a river from the open ocean. It is best to enter the river on a rising tide while the current is flowing inland. On an ebbing tide the prevailing wind opposes the current, creating dangerous waves at the river entrance. If unfortunate enough to go aground on route up the river, it is fairly easy to get off as the tide rises. In December 2001 we left Coffs Harbor in the afternoon, had a pleasant overnight sail under a full moon, and arrived at the entrance to the Camden Haven River around 8 the next morning. It was just before high tide, and we had a one-knot current helping us along as we motored up the channel, well marked by buoys, to the small town of Laurieton. Jim and Lynn, in the sailboat *Windchime*, left Coffs Harbor about the same time as we did, came up the river just behind us, and anchored near by. This was our third visit to Camden Haven, but a first visit for Jim and Lynn. For three days we shared with them the many delights of the small town of Laurieton.

It was perfect summer weather for climbing North Brother, the tree-covered mountain just west of town. It takes about two hours to climb, on a fairly steep, but well-maintained trail. We knew it was well worth the effort for the spectacular view from the top of the river winding its way out to the ocean. At the beginning of the trail we saw wonga pigeons calling with their deep woo, woo, woo. On the slopes of the mountain we saw colorful crimson rosellas, king parrots, and currawongs flitting among the trees. On shore, adjacent to the dinghy landing, the local yacht club maintains a restroom with a free shower for the use of cruisers, a great luxury. After our long hikes we always looked forward to a nice warm shower.

On our third day at Camden Haven we listened to the weather forecast on VHF in late afternoon and learned that a front had hit Sydney that afternoon with 50-knot winds. Storm warnings were out for the coast north of Sydney, forecast to reach our area by evening. When Jim and Lynn came back from their hike we told them about the storm forecast. Dan let out 50 more feet of chain to increase the scope to seven-to-one to make sure we didn't drag anchor. Soon after we finished dinner we saw dark menacing clouds building up over North Brother. It became a huge black cloud bank filling the sky above the mountain, with jagged bolts of lightning streaking from the clouds to the ground. The storm hit us at 8:30 pm, with powerful gusts of wind. One gust hit our boat on our starboard side, heeling our boat over so far that our scanner fell off the bunk onto the floor. Fortunately it wasn't damaged. It was really scary as the wind built up to 50 knots. After blowing hard for 45 minutes, the wind gradually eased up, and all was calm again. We congratulated ourselves that we hadn't dragged anchor. We went to bed in peace and quiet. At 10 pm, just as we were drifting off to sleep we heard Jim's voice calling "*Shaula, Shaula*, this is *Windchime*. We noticed your boat is facing in a different direction from all the other boats. Do you think you might be aground?" Dan called back, "I don't think so, but I'll check the depth." Dan turned on the depth sounder and was astounded to see a depth of 3.8 ft. The wind must have pushed the boat onto the shallow sandbank on the other side of the river because of the extra chain Dan had let out. All the other boats were afloat and lying to the current. We checked the tide tables. The tide would be rising for only the next 40 minutes. We would be high and dry if we didn't get off before the tide began to drop. Jim offered to help us kedge off, which we gladly accepted. Dan untied our Danforth anchor from the foredeck, attached 50 feet of chain and 200 feet of rope to it, then rowed over to *Windchime* and picked up Jim. They lowered the anchor into the dinghy, and rowed out 100 feet ahead of our boat to drop it in deep water. They returned to *Shaula* and Jim began the strenuous job of cranking in the rope with the anchor windlass while Dan tailed. Slowly, inch-by-inch, they dragged our eight-ton boat through the sand into the deep channel in the middle of the river. We got off just before high tide. By the time Dan pulled up the kedge anchor, re-anchored in deeper water, and took Jim back to his boat it was midnight. It was a great relief to finally be able to get some sleep in the calm waters.

This was the first time we had ever kedged the boat off, a technique used for centuries to pull boats off after going aground. We had read how to do it, but had never needed to use the technique. With Jim's help it worked surprisingly well. We were grateful to Jim and Lynn for noticing we were aground and for Jim helping us kedge off. If we hadn't got off the sandbank before the height of the tide we would have spent a very uncomfortable night in the boat. If we had gone aground on the coast, with waves pounding the boat on rocks, it could be catastrophic, damaging or destroying the boat. It is embarrassing to go aground while securely at anchor because of being overly cautious, but it didn't harm the boat, just our egos. We hoped we would never again be so careless as to go aground while at anchor.

## A Visit to Pretty Beach

A year after arriving in Australia, we sailed Broken Bay, an extensive cruising area just 30 miles north of Sydney. We anchored in Refuge Bay, a large well-protected harbor in Ku-ring-gai Chase National Park. The park has many beautiful inlets surrounded by eucalyptus trees. There are hiking trails all along the shoreline. As soon as we were anchored a beautiful wooden sailboat, named *Mother Goose*, motored past us. We chatted with the owners briefly, and they invited us to come over for coffee and to see their boat the next day. Thus we met Ken and Barb Beashel of Broken Bay. We exchanged stories of cruising in the South Pacific. We were impressed that Ken had built the boat and had sailed with their whole family across the Pacific to the U.S. and back, when the children were young. Our cruising in a boat that we bought, and sailing downwind from the U.S. to Australia seemed a much more modest accomplishment. The Beashels owned a boatyard not far from Refuge Bay, and their sons were involved in running the yard. Ken had been a racing sailor in his younger days, and their sons were also racers, one son being on the Olympic racing team for Star class boats. They were a true boating family, and we greatly admired their extraordinary accomplishments in boatbuilding, racing and cruising.

Our next visit to Broken Bay was the year after we traveled up the Queensland coast. We had hit the rocky bottom at the entrance to the Pancake Creek anchorage, and it was time to check out the damage to the bottom of our boat. We also needed to do some minor repairs and apply antifouling, so we decided to haul out at the Beashel's boatyard in Elvina Bay. It was a well-run boatyard where we could be hauled out on railroad tracks and work on the boat ourselves. While we waited for the boat that was in the yard to be finished and re-launched we had several rainy days so we stayed on

*Shaula* most of the time. There was little to do except read and play Scrabble. Finally it cleared up enough that we could go ashore to hike along the trail to the next bay, Lovett Bay. At last we could stretch our legs and have a walk on a trail through the eucalyptus forest. It started to rain again as soon as we returned, so we began another game of Scrabble on our small board set in the middle of the cabin sole. About an hour into the game Dan noticed an itch on his foot, and discovered a leech poking out from under the strap of his Teva sandals. Dan took his shoes off, and found four more leeches that were hidden by the straps, all swollen from having sucked his blood. I took off my Tevas and found I had been attacked also. We pulled the leeches off one by one, deposited them in a plastic bag and threw it in the garbage. An hour after we resumed our Scrabble game we noticed small pools of blood on the floor beside the Scrabble board. Unbeknownst to us the leeches had injected a blood thinner before they began sucking our blood, and we were both still bleeding for several hours after we removed them. Our feet itched for days wherever leeches had attached. Even though they pose no health threat, leeches are certainly the most despised creatures we've encountered on Australian soil.

The following year when we sailed to Broken Bay we visited Ken and Barb at their boatyard. They were going up the river to their vacation home at Pretty Beach, and they invited us to visit them there. The idea was appealing. It is a part of Broken Bay we had never seen, and we love exploring new places.

They told us to arrive at 9 am at high tide. It is important to navigate the channel on a rising tide, because the current would be with us and if we went aground it would be easy to get off as the tide rose. We left the next morning around 7, and wound our way up the channel, which was well marked by buoys on either side. The depths were mostly 10 to 12 feet, and as our draft was five feet we had no trouble finding our way up and avoiding the sandbanks on either side. It was only when we reached the anchorage in front of the Beashel's house that we worried—the depth was less than 10 feet, and we had to be sure the tide wouldn't drop more than 5 feet. *Mother Goose* was on a mooring near by. As soon as we were anchored Ken rowed out to our boat from the beachfront and invited us ashore to have coffee on the patio in front of their house. We had a great visit, and admired their beautiful house on Pretty Beach. They seemed to have an ideal, peaceful life, living on the water away from roads and traffic. In the afternoon we hiked the many trails around the shore to Hardy Bay, just adjacent to Pretty Beach. That evening we joined the Beashels for dinner and again enjoyed their wonderful company.

We planned to leave on the rising tide the next morning, hoping to make it down the channel to Broken Bay before the top of the tide. Before we left we decided to motor around to explore Hardy Bay. We raised anchor and motored out in that direction when I noticed some mangroves just off our starboard side. I told Dan that we were awfully close to the mangroves, which meant shallow water, but he insisted there was plenty of depth. In only seconds our boat came to a sudden halt. We had struck the bottom on a sandy spit extending out from shore. Now we were in serious trouble. As the tide continued to rise we would be pushed by the current into shallower water. If we were still aground at high tide we would be stranded for the rest of the day. There was only one solution—untie our spare anchor from the stern of the boat, launch the dinghy off the cabin top, and row the anchor out at a 45-degree angle from the bow of the boat into deeper water. This would hold the boat from being pushed backwards, and allow us to kedge off the sandbank when the tide rose sufficiently that our keel was off the ground. It took a nerve-wracking 45 minutes for the tide to rise. Once the boat was afloat we cranked in the line and used the engine to help pull the boat off the sandbank. Then we had to retrieve the anchor, stow it, and proceed down the channel. By this time we had used up the extra time we had before high tide, so we had to skip exploring Hardy Bay, and immediately head down the channel to Broken Bay. We made it down just before high tide.

It is humiliating to run aground due to carelessness, but it is a well-known saying among cruising sailors that if you never leave the marina you'll never go aground. All in all it was a great excursion to Pretty Bay, in spite of our unexpected grounding.

## Violin Disaster

We returned to our boat in Lake Macquarie in early December 2002, after a six-month visit to the States to help Dan's parents and to visit Seattle for our medical checkups. I had taken my violin with me, as usual, in order to practice when I had the time, and to play chamber music with friends in Seattle.

That summer fires broke out around Lake Macquarie. Bruce and Thelma, on *Tui of Opua*, were watching our boat, and were ready to motor her out of the marina to a safer part of the lake if the flames and smoke approached within 100 yards. When we arrived back the fires were out, but it was exceedingly hot. We had several days of 95-degree temperatures followed by several days of torrential rain and wind. The west winds were blowing from the hot interior desert, making it extremely uncomfortable. We took walks on paths around the lake, but were confined to the boat when we had downpours.

The first week we were back I got my violin out and practiced a couple of times. Then a week later when I opened the case to get out my violin I had a terrible shock. My violin was in three pieces. The fingerboard, the long ebony board attached to the neck, had fallen off, and the neck had broken off from the body of the violin. It was devastating. I realized what had happened. My violin was in its usual place in the quarterberth with the sun beating down on the cockpit seat just above it. In the hot summer weather the temperature had probably reached 110 degrees or more in that enclosed space. The heat and high humidity had melted the glue that holds the violin together. After the fingerboard detached the block holding the neck to the body of the violin was ripped apart by the tension in the four strings, causing the neck to break off. Over the many years of cruising with my violin I had had the fingerboard fall off, the heavy gut holding the tailpiece onto the tail peg break, and the sounding post fall over, but these were all easily fixed. This damage was far worse than these mishaps. It would mean a major repair by a violinmaker.

Unfortunately in Australia everything shuts down for about a month for the Christmas and New Years holidays. There was a small shop, Hunter Valley Violins, located close by on the north side of Lake Macquarie, but it was closed until January 15. There was nothing to do but wait. We enjoyed Christmas and New Years with cruising friends and I tried to forget about my broken violin. As soon as the shop re-opened I took my violin on a bus along the shore of the lake to the small shop. The owner examined it, and said he had seen many violins damaged by leaving them in a hot car, but never from leaving one in a boat. The top and bottom would have to be steamed off, and then re-glued, and the neck glued back on. It would cost $1,200. I asked what the violin was worth, and he said it was essentially worthless. There were dozens of new and used violins hanging along the walls of the shop, and he suggested I buy a used European violin for $2,600, or a new Chinese violin, which appeared to be mass produced, for $450. I couldn't imagine abandoning my old violin, which I had had for over 50 years. My violin was like a beloved pet to me. It was as if I had brought an injured cat to the vet, and was told to abandon it and buy a new one. I left the shop, saying I would have to think it over.

Fortunately another violin repair shop, the Sydney String Center, opened a week later. When I called the owner he was much more encouraging. He thought it would cost about $600. The next day I took a two-hour trip by train, then a short bus trip to his shop. The owner, David, was very helpful. He thought it would take about four weeks to repair. I felt I was leaving it in good hands, as this was the shop to which Sydney Symphony Orchestra players came to have their instruments repaired. As promised, he was able to repair it in four weeks. I took another train and bus ride to Sydney to pick it up. Because it had been six weeks after it was damaged before I brought it in for repair David was unable to bend the two curved sides completely to the neck. He had pressed them down as far as he could, then put in spacers to fill the gap. He had stained the spacers with a dark stain to make them blend in. I was extremely grateful that he was able to repair it. The cost was $565. My violin was whole again, and to me it sounded better than ever.

This violin had been a Christmas present from my parents when I was 15. It had belonged to one of my violin teacher's older students who had to give up playing when she developed painful arthritis in her hands. Having a violin of my own, not passed down from an older sister, was something special. I had enjoyed playing it in school orchestras, community orchestras, music camps, and chamber music groups. After a gap of 22 years, while I worked and attended graduate school, I took it up again in Seattle. In an Experimental College course called "Closet Violinists" I found many other violinists who had taken up playing again after a long absence. From there I graduated into playing in a small string orchestra. Soon I was playing in a community orchestra and got together weekly with other string players to play trios and quartets. I took up playing the violin again with great enthusiasm.

When we went cruising across the Pacific it seemed natural to take my violin along. Although it was impossible to find other violinists on cruising boats, lots of folks brought along other instruments, such as guitars, harmonicas, banjos, ukuleles, recorders, accordions, flutes, electronic piano keyboards, and drums. Whenever possible we got together to play music on someone's boat or on shore. The pleasure I got from being able to play music wherever we went far outweighed the risk of damaging the violin.

## Boomerang Trip

In 2003 we planned to sail from Coffs Harbor to Vanuatu, stopping in New Caledonia. The passage to New Caledonia is about 700 miles, which should take about a week. Coffs Harbor is our preferred port of departure for New Caledonia because the customs office is adjacent to the marina, making it easy to check out before leaving. Cruisers are required by customs to leave within 24 hours of checking out. If a problem comes up and they are unable to leave, they can check back in and wait for another weather opening before checking out again. Leaving from Sydney is much less convenient, requiring taking an hour trip by bus or train to the airport to check out.

In May we began sailing up the coast from Sydney to Coffs Harbor. Our first stop was Newcastle, where we stayed in the new marina. While at the marina we could step ashore onto a dock to walk on shore to the shops, and to take hot showers. We didn't want to pay $30 to stay another night so we left the marina the next day to anchor out in the nearby Hunter River for free. The next morning we planned to raise anchor and sail to our next anchorage, Port Stephens. After I cranked up the anchor using the windlass, Dan went to the foredeck to stow the anchor on the bow roller. Then I began motoring down the river. However, I noticed by watching the trees go by on shore that we were moving at a nice pace, but backwards up the river towards the bridge. For a heart-stopping moment I thought our boat was going to be swept upriver by the current and onto the concrete bridge supports. I yelled to Dan: "We're going backwards—I can't get it into forward gear." He calmly replied, "Let's drop the anchor and I'll check out the problem." He quickly dropped the anchor, and the boat stopped moving towards the bridge. Once we were re-anchored Dan crawled into the engine compartment to check the transmission. Fortunately it was only a broken cable that connects the gearshift to the transmission. It broke off about an inch from where it was connected to the gearshift. Dan was able to re-connect the shortened cable and we had forward, neutral, and reverse again. Once again we raised the anchor, motored down the river and began sailing to Port Stephens. When the wind died later in the afternoon, we started motoring again, and immediately heard a loud squeal from the engine compartment. Dan recognized the noise as that made by the engine's fresh water pump—the bearings were worn out. So we continued on to the anchorage, sailing as much as possible. Now, in addition to ordering a new transmission cable, Dan would have to replace the water pump with our spare. So far we had no serious problems that would keep us from departing for New Caledonia.

We were sailing in company with two couples with whom we had shared an anchorage in Sydney Harbor—Frank and Therese from Scotland, and Joop and Annaliese from Holland. Although we each went at our own pace, we enjoyed getting together at anchorages along the way either to have happy hour or dinner on each other's boats or to go ashore for meals. We shared a wonderful dinner at an Indian restaurant at Port Stephens, and a dinner at the Thai restaurant at Camden Haven. We also climbed North Brother while there. From Camden Haven we had an overnight sail to Coffs Harbor.

At Coffs Harbor we began to look for a weather window for our passage to New Caledonia. From the marina there is a nice walk along a creek into town, where the public library has free Internet. We used a new website called Buoy Weather to get seven-day forecasts for the area between Coffs Harbor and New Caledonia. For the first several weeks there were no favorable forecasts, so we spent our time provisioning with staple food items, and buying lots of trading items, such as fish hooks, bar soap, pens and pencils, thread, and needles for Vanuatu. To make room for all the provisions and other items we bought we had to get rid of 150 pocket books that we had accumulated while cruising along the Australian coast. When we got together with friends for happy hour we brought bags of books we had read or didn't particularly want to read, and gave them away. Ones that we couldn't give away we took to the laundry room at the marina, where books are exchanged among cruisers. Finally we cleared out enough space so that we could store the provisions for five or six months for our trip. Dan spent much of the time replacing all the valves in the head (toilet), and replacing our fresh water pump and the broken transmission cable.

Finally we found a weather window starting June 4$^{th}$. The day before departure we bought the fresh food we would need for the passage, stowed everything, got out the charts we would need, and went to the customs office to check out and have our passports stamped. The winds were in the right direction and we were happily on our way that afternoon, sailing northeast, with the winds behind us.

Our passage went well until two days out. Then on the morning ham radio net our friends in Australia, who had been listening to the high seas weather forecast, mentioned that a cyclone had formed just north of New Caledonia. The cyclone would be directly in our path as we headed to New Caledonia. We had carefully avoided the possibility of sailing into a cyclone by never sailing in the tropics during the cyclone season. This, however, was an out-of-season cyclone, and was a real concern. We continued on until afternoon, when we heard on the weather report that the cyclone had intensified and was heading southwest—directly towards us. It was still 500 miles away, but we were very uneasy. We looked at each other with stricken faces, and Dan said, "I don't think we want to gamble with our lives by continuing to sail towards a cyclone." I readily agreed. This would be the worst weather we had ever encountered in our 10-year cruise since we left Seattle. We quickly turned the boat around to head back towards Coffs Harbor. We were 186 miles off the coast, and figured it would take two days to return to port where we could wait out the cyclone before departing again.

While underway we had noticed a streak of water running across the cabin floor when we were heeled over on a port tack, and figured out that we had a leak in the seacock for the sink drain. We had also noticed that our batteries were not functioning properly. They ran down while we had the navigation lights and electronic equipment on overnight, and could not be charged up completely when we ran the engine. We were hoping to fix these problems when we returned to Coffs Harbor.

On the trip back we motored the rest of that day, but by the next morning the wind had picked up and was blowing 20 to 25 knots from the south. With an adverse current our speed had dropped to less than three knots. We tried sailing that afternoon from 3 to 11 pm, but by midnight the wind had increased to 20 to 30 knots, right on our nose. We decided to hove to in the gale. The high seas weather reported that the gale area extended for 600 miles north and south and 300 miles east to west, so there was no way to motor out of it. It was a miserable time, with the boat drifting slowly at about one knot with 10-foot waves slapping repeatedly against the hull. We still had to keep watches, looking around every 10 minutes to be sure we were not in the path of any ships. We tried to sleep during our off watches, but the motion made that almost impossible for me. I wasn't cooking anything, so we just ate apples, bananas, crackers, cookies, and anything else we could scrounge. Finally, after 36 hours, the winds abated to 20 to 25 knots from the south-southeast, and we could sail west. Dan took the first watch, and I went down below to crawl into my bunk and try to sleep. We were heeled over to starboard, and as my bunk was on the port side (the high side) I had to fasten my lee cloth to keep from sliding out of my berth. As I was tying the knot in the lee cloth line, the boat gave a lurch, and I slid out, landing on the floor sitting face forward. Before I could pick myself up the boat gave another lurch, throwing my head backwards against the corner of the small shelf on bulkhead above me. I heard a loud clunk as my head hit the bulkhead, and felt pain in my scalp. Then blood began pouring over the side of my skull, into my ear, down my chin and onto the jacket of my foul weather gear. Although my scalp was bleeding a lot, I didn't think the wound was deep, so I just used paper towels to sop up the blood. For the rest of the trip I slept in Dan's bunk on the low side of the boat. The blood dried in a long streak on my scalp and hair, but when I tried to soak it off a few hours later it started bleeding again, so I left it alone.

The winds were calmer the next day, and we decided to motor on into Coffs Harbor in order to get there before sunset. When we got close we called the marina on the VHF to tell them we were returning and needed a slip. We were assigned a slip next to our friends from Scotland. As we approached the dock we saw four of our friends standing on the dock ready to take our lines—Frank, Therese, Joop, and Annaliese—plus the customs officer. It was wonderful to have such a warm welcome. After we filled out some forms for customs, Therese and Frank invited us for dinner on their boat along with Joop and Annaliese. No one we talked to could believe we had been in a gale for two days, because at Coffs Harbor they had had beautiful, sunny weather for the last six days. We had a great time over dinner telling the story of our passage. Our fellow cruisers all thought we had made the right decision returning to Australia.

We learned that when cruisers have to abort their passage and return to Australia it is called a boomerang trip. We had made a six-day passage to nowhere, but were very happy to be back in port. Even though we had been caught in a gale, we at least had avoided sailing into a cyclone.

## Crossing the Bar

After our three-month winter cruise to Vanuatu in 2003 we made landfall in Bundaberg, Queensland in October. It was a week before the arrival of cruisers in the Port to Port rally from Port Vila, Vanuatu, to the Port of Bundaberg. The Bundaberg marina was fairly new and trying to attract people to moor their boats there by offering free van rides into town for shopping, and free Friday night barbecues of hamburgers or fresh fish with bread, and a variety of salads. Cruisers are nearly all on a tight budget, so free events are very well attended. If we didn't get in line as soon as the food was served it was all gone. We enjoyed watching the boats arrive from Vanuatu. Everyone was in a festive mood, having just made an eight-to-10-day passage. The organizers of the rally had a week of celebrations, dinners, and prizes planned for the 30 or so boats that participated in the rally, and we were invited to take part in some of the events. Then it was time to head south to New South Wales for the spring and summer season to avoid the hot sticky weather and possible cyclones on the Queensland coast in the summer.

After leaving Bundaberg we had to negotiate the long, narrow, and shallow channel between Fraser Island and the mainland, known as Great Sandy Strait. Even though the channel is marked by red and green buoys it is a nerve-wracking experience, because any deviation from the middle of the channel can land a boat on a shallow sand bank. We went through on a rising tide so that in case we grounded we would be able to float off when the tide rose. With the wind behind us we were able to sail through the channel instead of motoring. We tried to stay in the center of the channel, always keeping an eye on the depth sounder. I called out the depths when the depth dipped below 10 feet: "8.5, eight, seven, 6.5, six." Now we were in trouble because we were on the edge of a sandbank and didn't know which direction to head to avoid going aground. The next reading was five feet (the same as our draft) and we hit ground. However the tide was rising, and we floated off in a few minutes. It took about three hours to reach the south end of Great Sandy Strait.

At the end of the channel we anchored in Pelican Bay to await a slack tide the next morning in order to cross the dreaded Wide Bay Bar. This is a much bigger challenge than Great Sandy Strait because there are no buoys to mark the deep channel. Instead boats are guided on a zigzag course by three GPS waypoints. The first leg is northeast, the second southeast, and the third directly south. That afternoon I called up the Tin Can Bay Coast Guard to get the latest waypoints, which can change from year to year with the shifting sands. I put the waypoints into a route on the GPS so that we could make it across the bar the next day. We had to get up in the dark the next morning, raise the anchor, and motor towards the first waypoint at the slack tide at 5 am. Everything was fine for the first hour as we were approaching the first waypoint, when we were jarred by the loud buzz of the engine alarm. Oh no, not again! We looked at the temperature gauge—it had risen above 180 degrees, so we turned off the engine to keep it from being destroyed. (When the temperature goes over 180 degrees the cooling water can evaporate, causing the engine to seize up.) Fortunately we had a light wind, so Dan quickly raised the main sail and staysail, and rolled out the jib. With all three sails set we could turn the boat around and sail slowly back through the channel to Pelican Bay to re-anchor.

We had had several scary experiences with our engine in the previous six months. Our engine overheated when leaving an anchorage on Ureparapara in Vanuatu, we had to sail in a narrow pass in the reef surrounding New Caledonia, and more recently in the long, dredged channel marked by tall piles at the entrance to the Port of Bundaberg. As long as we were motoring in smooth waters the engine behaved well, but when waters became choppy, we noticed that air would be sucked into the intake, and soon cooling water stopped coming out of the exhaust, and the engine overheated. The saving grace in a sailboat is that as long as there's wind we can sail instead of relying on the engine. We prefer sailing anyway, because it is much quieter than motoring.

After our aborted crossing of the Wide Bay Bar that morning we sat on our boat for the rest of the day at the anchorage at Pelican Bay contemplating the problem, trying to figure out why the engine was overheating. Dan thought that the problem might be the grate over the saltwater intake. The grate is a three-inch long rectangular metal fitting with narrow slots that covers the saltwater intake pipe. It prevents large pieces of seaweed or other debris from being sucked into the engine's saltwater cooling system. When we hauled the boat out the previous year to antifoul the bottom Dan had accidently put the grate on backward, so the narrow part faced forward. Possibly the narrow end of the grate was obstructing the intake. So after talking to other sailors on the radio, who said the grate wasn't really necessary, Dan dove over the side of the boat, unscrewed the screws that hold it on, and removed the grate. We were ready to try to cross the bar again.

The next morning we once again got up while it was still dark, raised the anchor and started motoring out to the first waypoint. We made it past the first waypoint, and were about in the middle of the bar when the loud engine alarm shrieked at us again, but unlike the previous morning there was not a breath of wind. Even though we immediately turned off the engine and raised the mainsail we were drifting with the current with no way to steer. I was really panicked. It was like coasting down a hill on a bicycle when suddenly both the brakes and steering fail. There were reefs on either side of the channel, which were invisible in the murky water. Our boat could easily drift onto a reef where we could be stranded for hours, with the boat pounding on the bottom. I asked Dan if I should call the coast guard. He reluctantly agreed. We had always been able to handle problems on our boat without asking for help. Now we had to swallow our pride and admit we couldn't solve this problem alone. I grabbed the VHF radio microphone and called: "Tin Can Bay Coast Guard, Tin Can Bay Coast Guard, this is *Shaula, Shaula*." The calm reply came back: "Yes, *Shaula*, how can we help you?" "Tin Can Bay Coast Guard, we're headed over the Wide Bay Bar and our engine has failed. We're unable to sail because there's no wind." "*Shaula*, do you need a tow?" "Yes, we're drifting in the channel without steerage way." "*Shaula*, we'll launch our boat and be out there to give you a tow. What is your current position?"

I gave him our GPS coordinates of latitude and longitude, and we waited, and waited. It was a long nail-biting 40 minutes as we drifted slowly towards the south end of the channel, being unable to steer away from danger. At last, with great relief, we saw the coast guard cutter with four men aboard coming down the channel towards us. Then we watched it speed right past, heading south down the channel. Now I really panicked. I quickly called Tin Can Bay Coast Guard again. They responded immediately, turned the boat around and returned to us, saying they thought we would be further along the channel than we were. Dan stood on the foredeck while one of the men tossed a long, heavy towline, which Dan attached to the two vertical bits on the front of the boat. We watched anxiously as the man on the coast guard boat accidently dropped his end of the rope, which he had to fish out of the water with his boat pole. Finally we were attached and they began to motor back through the channel towards the marina at Tin Can Bay. They seemed very professional at towing, motoring at a steady four knots, so as not to put too much stress on the attachment to the boat. While we were being towed Dan and I took turns steering our boat to keep it from slewing from side to side, and eating breakfast. It took about two hours to get to the marina, where they deftly pulled us to the visitor's dock and asked Dan to release the towline, so we could jump off and tie up. They were on their way back to the launching ramp before we could even thank them. We were so grateful for their help in saving our boat, a job that they do voluntarily, with no compensation from the government. I baked a batch of chocolate chip brownies that evening that we took over to the coast guard office the next day along with a $50 bill that we hoped would at least pay for the gas they used in the rescue.

After talking to a couple of mechanics at the marina and to a cruising friend on the ham radio net the next day, we concluded that the problem we were having with the engine was caused by our failing saltwater pump. Our friend said that ordinarily the saltwater pump can overcome some air being sucked into the intake, and will quickly recover to keep the engine from overheating. Our 15-year-old pump had been pumping less and less efficiently over the years, but we hadn't noticed because it had happened so gradually. The next day Dan ordered a new saltwater pump, which the dealer in New South Wales had in stock, at a cost of about $500. It came in the next day, and Dan immediately installed it. Now, when we started the engine, we were amazed to see saltwater gush out of the exhaust pipe within a few seconds, instead of the seven or eight minutes it had been taking, and at about twice the volume of our old pump. Our problem was solved, and a few days later we motored safely cross the bar with a great sigh of relief. It was humiliating having to call for help from the coast guard, but it was a wonderful feeling to know that there were men who would take time out of their busy lives at an instant's notice to help out sailors in distress. We had great respect for the Australian volunteer coast guard and their very efficient way of handling emergency situations.

## How I Became an Illegal Alien

In late November 2003, a month after we returned to Australia from Vanuatu, we sailed south from Bundaberg to Brisbane, and tied up at a marina in Manly. Although we had four-year multiple entry visas, due to expire in two more years, we decided to apply for four-year retirement visas. Many of our cruising friends had obtained these special visas, which have the advantage that we would not have to leave within six months of entering Australia each time we returned from overseas. We knew about the grueling application procedure, which includes paying a hefty application fee, fingerprinting, a police check in every country we had visited in the last 10 years, a chest X-ray and physical exam by an Australian doctor, proof of adequate health insurance, and proof of sufficient funds to support ourselves for four years. They make it tough, but we thought we could qualify.

I called the Brisbane Immigration Office to ask if we could pick up the application forms. The woman at the immigration office said that we could apply as long as we held valid Australian visas. She asked my name, and looked up my visa status in their computer, then asked where I was calling from. I said I was in Brisbane, and she said: "I don't find that you have a valid visa, and don't see any record that you left Australia from Coffs Harbor in July or that you returned to Bundaberg in October. You should come into the office with your passport to straighten this out."

This was strange. The customs officer in Coffs Harbor had stamped our passports with an immigration stamp and the date of departure in June, and again in July when we departed for the second time after our boomerang trip. When we arrived in Bundaberg from Vanuatu in October, the customs officer had stamped our passports with our arrival date. Our four-year multiple entry visas were glued into our passports. How had the immigration department lost this information? The immigration office is located in downtown Brisbane, very easy to reach by train from Manly. We caught the train the next morning with passports in hand.

After an hour wait, we came up to the desk, where the young man took our passports, and looked up our data in his computer. He seemed puzzled with the results. We told him about our boomerang trip from Coffs Harbor due to a cyclone, and explained that when we returned six days after departing my six months on my visa had expired. Dan was still within the six-month period of his visa because he had made an additional trip home in February for his father's memorial service. The customs officer at Coffs Harbor, who handles immigration, assured us on our return that we didn't have to worry about my being past the allowed six months; he would contact immigration and get a bridging visa for me. A few days later he told us that immigration had granted me a bridging visa until the end of June. When we were unable to fix all our boat problems by then, he was able to get a three-week extension of the bridging visa. We didn't have anything in writing, but we took the custom officer's word for it. The immigration officer was skeptical. Our departure date was not in his computer, therefore he didn't believe we had departed. After much consulting with his superiors, and much argument with us, he sent us up two stories to the 13$^{th}$ floor. When we arrived at this office we knew we were in trouble. The sign on the door read "Office of Compliance and Investigation of Unlawful Non-Citizens."

I told my story again to the immigration officer here. He explained to me that when I returned to Australia June 10$^{th}$, four days after my six-month visa period had expired, my four-year visa was automatically cancelled (unbeknownst to me). He said the customs officer in Bundaberg who stamped my passport when we entered in October should not have allowed me into the country without a valid visa. I was now an illegal alien, and had been one since I entered the country in October. We were devastated. We had been reading newspaper articles on refugees who had arrived by boat and tried to enter Australia without visas. They were put in detention centers in Australia, or turned back before reaching Australia and put in detention centers on Christmas Island. Some had arrived in leaky boats, and had drowned after they were turned away. The Australian Immigration Department does not treat illegal aliens kindly, to say the least. Compassion is not part of their job description.

Several hours of pleading were to no avail. I tried to remain calm and reasonable, although I felt anything but calm. I argued that I was innocent—the customs officer in Coffs Harbor, who is supposed to act as an immigration officer, didn't tell me that my visa was invalidated. The customs officer in Bundaberg let me into the country without a valid visa. Neither of them sent the dates of our departure or arrival to the immigration department. Why shouldn't they bear the responsibility for their mistakes? I was not told what a bridging visa was—a temporary visa to allow me to stay in the country until I could leave and obtain a new visa. I could have easily got a new visa when we were in Vanuatu. The immigration officers were very firm. Their answer was: "If you don't ask, we won't tell." Because I didn't know what questions to ask I didn't get the information from them as to how to comply with their rules. We were defeated. I was being treated like a criminal. I had never broken a law, except for a couple of parking violations in Seattle, and even then it wasn't intentional. We went back to the 11$^{th}$ floor to talk to the young immigration officer again. He told me I was being deported, and gave me a bridging visa until December 25$^{th}$. I could only apply for a new visa from outside the country, but there was no guarantee that it would be granted. He said he was being lenient—he could have deported me with a three-year exclusionary period before I could apply for a new visitor's visa. This development put a serious crimp in our plans to apply for retirement visas while in Brisbane.

Being denied a visa would be disastrous to our plans. Dan would have to find crew to help him sail our boat to New Zealand if we wanted to resume our cruising. However, when we left New Zealand five years before we swore we would never sail there again. Being knocked down by a rogue wave on the way to New Zealand had cured us of the desire to sail in those treacherous waters. I was very discouraged and depressed.

Two days later at a Thanksgiving potluck at the marina we told our story to our cruising friends. Most were appalled at the way we had been treated. Many Australians thought we should protest against deportation, and try to get the help of the local parliamentarian. Some American friends had had problems with Australian immigration, and were very sympathetic. We decided not to fight it, but just make the best of a bad situation. We would fly to New Zealand for a four-week holiday, where we could apply for a new visa. While waiting for it to be processed we would visit parts of the North Island that we had missed during our two and a half years cruising along the New Zealand coast. It would be a relief to get away from Australia for a while. We were looking forward to a Christmas holiday with the friendly New Zealanders.

## A Trip to New Zealand

We had planned to travel inland in Australia in February of 2003—to visit Melbourne and Tasmania by train, ferry and bus—but our plans had to be put on hold when immigration told me my visa had been cancelled and I had to leave the country in order to apply for a new one. Immigration gave me a bridging visa until December 25$^{th}$, so immediately after Thanksgiving we started making plans for a trip to New Zealand. We chose to fly to New Zealand rather than the United States because tickets were much cheaper, and it is only a three-hour flight. When I called the immigration office in Auckland I found out that it would take up to two weeks to issue a new visa, and they would be closed four days for Christmas and three days for New Years holidays. We decided to leave by December 6$^{th}$ and return January 6$^{th}$ so that New Zealand immigration would have a couple of weeks to work on my visa before the holidays began and we would have time to do some sightseeing. I booked two return airline tickets for those dates.

A few days before we were to leave we went out to dinner with a group of cruising friends, and learned from a Canadian couple that I wouldn't be able to board the plane in Brisbane because I had a return ticket, but didn't have a visa for returning to Australia. Jack and Norma had unknowingly overstayed the six months on their visa before they flew home to Canada, and were stopped from boarding their flight back to Australia. They had to call Australian immigration from the airport in Vancouver and argue and beg for permission to board their flight. So the next day I called the airlines, and they confirmed that I wouldn't be able to board the plane with a round trip ticket without having a valid visa for the return. The only way to solve this problem was to purchase a fully refundable ticket to the United States from New Zealand, at an exorbitant price of $3,300!

Now we had a day to pack, close up the boat, and print out the visa application forms I would need in New Zealand. We had little time to actually plan our travels once there.

It was a pleasant flight to Auckland. We had booked inexpensive accommodation for the first three days. We spent the next day, Sunday, filling out the eight-page application form, which required listing the dates of travel for all the countries we had visited during the last 10 years and a long explanation of why I needed a new visa. We arrived at the immigration office at 9 the next morning. While waiting for my number to come up we talked to a young Chinese student who had lost her visa when her purse was stolen in Sydney. We sympathized with her. It seemed absurd that she had to travel to New Zealand to apply for a replacement visa. When I told the immigration officer that my visa had been invalidated, she said this usually means there is an automatic three-year exclusionary period. Then, when I told her the complicated story of returning to Australia after the six months allowed on my visa to avoid sailing into a cyclone, she was annoyed. She couldn't understand why the immigration department in Australia couldn't handle my case there. She said she'd work on it and let me know the outcome by email. Now we could travel the North Island of New Zealand while waiting to know the outcome of my application. My ticket from New Zealand to the United States had served its purpose of allowing me to board the plane, so I called Qantas and cancelled it immediately. We felt we were in limbo, separated from our home and cruising life of the last 10 years, not knowing what the future held.

For the next few days we took our minds off my visa problem by visiting with New Zealand cruising friends we had sailed with along the coast of the North Island five years previously. Brian and Sue had retired early and lived on their boat, *Waitoa*. They were excellent sailors, guides, and fishermen. We had shared many anchorages and seafood dinners with them. We really enjoyed visiting them again. We also met up with Stan and Marge whom we hadn't seen since our first visit to Vanuatu in 1998. When we called them they invited us for dinner with their many relatives, and showed us their new three-storey home that they now shared with their daughter, son-in-law, granddaughter and great granddaughter—a wonderful family.

We continued north from there to visit the Waipoua Forest, where the few remaining kauri trees grow. Kauri trees were treasured for their strong and durable wood for building boats and furniture, but were nearly wiped out by the early settlers of New Zealand. We took several hikes to see these huge trees, some over a thousand years old, and so rare that they had individual names. The nearby Kauri Museum showed the many uses of this beautiful wood, as well as the gum, known as amber. We had taken a tour bus, and had made reservations to stay in the small town so we could spend all afternoon enjoying the museum. We continued the next day on a bus to Cape Reinga, the northern tip of New Zealand. We saw the famous 90-mile beach, and the steep sand dunes, where we delighted in watching the tourists zoom down the sides of the hills at great speed on boards resembling snowboards. As we were in our late 60s, we decided to forego that experience. We hiked to the Cape Reinga lighthouse, the huge beacon that welcomes sailors to New Zealand if the visibility permits it. The next day we stayed in a small hostel at the Bay of Islands, where we had sailed when we first entered New Zealand. We took an hour hike to the boatyard at Opua, where we had replaced our boom eight years before after our knockdown on the way to New Zealand.

Our mood improved greatly when, just before Christmas, I received an email saying my new visa had been granted. We traveled back to Auckland to pick it up before the office closed for the Christmas holidays. It was a very welcome Christmas present. The best part was they granted me a visa identical to the one that immigration had invalidated. Our mission accomplished we could relax and enjoy the rest of our stay. We did some sightseeing around Auckland, visiting Rangitoto, a steep volcanic island, which can only be reached by ferry. We hiked to the top of the volcano, through the largest pohutukawa forest in the world. It had been restored by ridding the island of opossums and wallabies that had nearly destroyed the trees. The trees are known as Christmas trees, as they are filled with beautiful red blossoms at this time of the year. During our stay in Auckland we had dinner at a Thai restaurant with old cruising friends who had settled in Auckland to work in computer technology. On Christmas day the buses were free, so we took an hour bus trip to the outskirts of Auckland to see the botanical gardens, with all the trees that are unique to New Zealand. While in Auckland we indulged ourselves with seeing several movies and eating at different ethnic restaurants.

From Auckland we took a train to Waitomo to see the famous glow-worm caves. After seeing the caverns filled with colorful stalactites and stalagmites we boarded a flat-bottomed boat to glide through the dark waters, propelled by the guide pulling on overhead wires. The boat entered a large grotto where thousands of gnat larvae with tails lit like fireflies hang from the ceiling, resembling distant stars. Two of the couples had brought their young children along on the boat. One of the children wailed continuously, probably out of fear of the water or the dark, spoiling the effect for everyone else. When we got to the end of the ride the two couples with children disembarked and departed. Our guide then took the rest of us back to the grotto to see the glowworms again for free. It was a much more memorable experience in the silence deep inside the cave.

From Waitomo we took a three-hour train trip south to Tongariro National Park. We stayed at a backpacker's hostel full of young people. Every morning a bus would pick up hikers who wanted to take the 12-mile hike known as Tongariro Crossing. The driver would drop them off at the start of the trail, and would return in late afternoon to pick them up at the end of the trail. The driver would let us stay on the bus, and drop us off at the visitor's center. Instead of taking the exhausting 12-mile hike across the base of the volcano in the heat of summer, we would have our morning coffee, and then take a couple of short hikes along the streams and through forests in this unusual national park. We enjoyed the slow seven-hour train trip back to Auckland. The tracks have been welded together to make a smoother ride, but on hot days such as this one, the rails tend to buckle. Our train had to travel at less than 25 miles an hour to prevent being derailed.

The friendliness of the people of New Zealand, and the beauty of the scenery on the North Island made up for the unpleasant experience of being deported. We would not have re-visited New Zealand had we not been forced into it by the ridiculous policies of Australian immigration. It was an opportunity to see much more of New Zealand than we had seen in two years by boat.

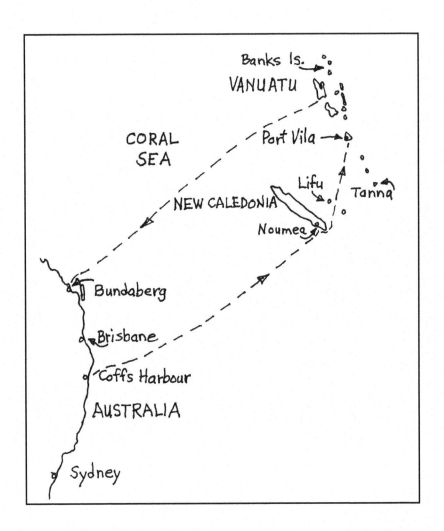

Australia to New Caledonia and Vanuatu

# 10 VANUATU AND NEW CALEDONIA

## Emae Island

On our second visit to Vanuatu in 2000 we made landfall in Port Vila, the capital city, and enjoyed the delights of the big city. In addition to shopping at the supermarkets and the public market, we ate lunches and dinners ashore at several cafes and restaurants, and had potlucks aboard boats with other cruisers in the anchorage. Towards the end of our stay we joined in the celebration of Vanuatu's 20$^{th}$ anniversary of independence. The Vanuatu people know how to celebrate, with singing and dancing every night on the seafront stage near the anchorage. We attended many of the events in the evening, but always left after a couple of hours to go back to the boat to sleep. This wasn't easy, as the singing and playing of string bands lasted all night, every night for a week, with loud speakers blaring out over the bay where we were anchored. By the end of the week we were exhausted.

Six weeks after arriving in Port Vila we decided to head northeast to the small island of Emae. The wind cooperated with a nice breeze of 15 to 20 knots and we made the 44-mile trip in only seven hours. Two boats of friends that we knew, *Anwagomi* and *Rassamond*, had arrived the day before, and reported the place was calm and beautiful. Another boat, *Kapilua II*, arrived just after us. We motored around the coral-strewn bay and found a patch of sand where we could drop our anchor. After lunch, Pony, from *Rassamond*, came over in his dinghy to take us and Len and Pam from *Kapilua II* to shore to ask for permission to snorkel on the nearby reef. We wandered down the path and ran into Adrian, who owned the beach where we had landed. He told us that the chief was in the village on the other side of the island, but it was all right for us to snorkel. He said there were sharks, but they weren't aggressive. When we returned to our boats Pony and Sylvia, and Len and Pam got in their dinghies and motored over to a large coral reef, and spent the afternoon snorkeling in the warm clear water. The rest of us, worried about sharks, decided to stay on our boats. The snorkelers came back disappointed, saying the coral was mostly dead, and they saw very few fish. It is probable that a cyclone had devastated the coral, leaving mostly coral rubble. It might take years for it to re-establish. Pony and Sylvia invited us and the crew from *Anwagomi* and *Kapalua II* over to their boat for happy hour later that afternoon, and we had some wonderful snacks of paté on crackers, fish dip, and popcorn. We were all enjoying the peace and beauty of this anchorage, away from the noisy celebrations of Port Vila.

The next morning Dan and I and Gordon, from the boat *Anwagomi*, went ashore on the island and took a walk along the trail towards the north shore. After 20 minutes we met Charles Meto at the small village of Matangi. He told us that about 100 people lived in the village—a welcome change from Port Vila with its population of 20,000. After we introduced ourselves he welcomed us to his village, and brought us over to a large pamplemousse tree at the edge of the village. Pamplemousse is our favorite fruit that we had enjoyed since we first sailed to French Polynesia 11 years before. Charles asked a half dozen children to pick some of these giant, green grapefruit for us. The children climbed up the tree, picked the large fruit and tossed them down to the ground. Charles retrieved them and filled up a large flour sack with pamplemousse. We had balloons in our packs to give the children who had picked the fruit for us. When we asked if we could pay for the fruit, Charles said no, but he needed fishing line. We promised to return the next day to return the flour sack and bring some fishing line. That afternoon we divided the pamplemousse among the four boats at anchor, and had happy hour on *Kapilua II* where we shared our stories of visiting the Emae islanders. After dinner we dug into our supply of trading items and gathered things we thought the villagers could use—a jar of rice, a jar of laundry detergent, two T-shirts, and a few magazines.

The next morning Dan and I went ashore with Gordon, who added a large reel of fishing line, fish hooks, a T-shirt, a notebook and pencils to our contributions. Gordon was on his way to another village to the north that he had visited two days before to help repair their broken generator. Dan and I headed for Matangi village where we met Charles again and gave him the flour sack full of trade items. He introduced us to his sister, Josephine, who told us there would be a church service at 10 am, and we were welcome to attend. While waiting for the service to begin we sat at a long table on Charles's porch to chat. The church bells chimed every 15 minutes, but it wasn't till the third time that everyone from the village began filing into the large building. Meanwhile, Josephine had changed into her Sunday best dress, as had the other village women. She handed us two bibles, one in English, the other in Bislama, a Pidgin English that is the common language of Vanuatu. We felt out of place, wearing our usual garb of shorts, T-shirts and sandals, but it didn't seem to matter to the villagers. We hadn't even realized it was Sunday when we went ashore, much less thought we would be attending church. We sat on a pew near the front of the chapel and listened to some beautiful hymn singing and a sermon, all in Bislama. The lay minister welcomed us as visitors, and asked us to join the reception line after the service. We stood next to the minister and other officials of the church and shook everyone's hand as they departed.

Josephine took us on a walk through the village and down to the beach, where we met some of the children of the village. She told us she had a cousin in Port Vila who was director of fisheries. He traveled all over the region as part of his job. When she was a teenager she lived in Port Vila to babysit for her cousin's children. She much preferred living at Matangi, which is quiet and peaceful, and less expensive than Port Vila. She invited us for lunch, which turned out to be a huge feast, consisting of three kinds of laplap (taro and bananas, yams topped with island spinach, and manioc topped with chicken), canned sardines, and pamplemousse and papaya slices for dessert. The table was beautifully decorated with colorful branches of bougainvillea stuck in pamplemousse as a vase. We felt like royal guests. Josephine introduced us to her father and her brother Charles's wife and young daughter. Josephine explained that on Saturday the men in the village hunt wild pigs, while Sunday was a day of rest. After attending church they have a big meal at noon, rest in the afternoon, and enjoy leftover laplap in the evening. Charles told us this was an historic occasion—the first time cruisers had attended a church service in their village. As we left Charles insisted on giving us a dozen more pamplemousse, which we greatly appreciated.

This was the first time we had attended a church service in Vanuatu. We had planned to take a walk on Emae Island just to stretch our legs, and we ended up being treated like honored guests at Sunday lunch in a small village. The best experiences we have had when cruising seem to always come from visiting small islands off the beaten track such as Emae Island.

## Lamen Bay

After leaving Emae Island we headed for Lamen Bay on Epi Island. The winds were from the southeast, making a very pleasant downwind sail to Epi, north of the capital, Port Vila. As we approached the anchorage, we were excited to find we had caught a huge fish on the line we dragged behind the boat. Dan pulled it in and laid it on the side deck—a 50-inch wahoo, one of the tastiest ocean fish. However, we had a problem—this was way too much fish for us to eat, and we had no refrigeration. With ten boats at anchor we decided it would be possible to cut the fish up, keep enough for a meal for ourselves, and give away the extra. Dan filleted it and cut one fillet into four portions, one of which we kept for our dinner. We rowed around the anchorage offering the other three wahoo pieces to cruisers anchored nearby. Amazingly, the first two boats we visited didn't want any more fish, but we finally found couples on three boats who gladly accepted our offer. We took the other large fillet ashore to give to Tasso. We had visited Lamen Bay two years earlier, and knew Tasso had a freezer on shore for food he serves at his beachfront restaurant. Tasso was delighted with the fish fillet, which he immediately put in his freezer. He remembered us from our last visit, when we had eaten a very nice dinner at his restaurant and had a long conversation with him afterwards. That evening on the boat I poached our piece of the wahoo in water and sherry and served it with potatoes and tomato slices—a welcome change from our usual vegetarian fare.

Tasso owns the only restaurant and resort in the small village of Lamen Bay where he lives with his wife and five children. He has built several thatched bungalows along the beach with an open-air restaurant at one end. Several years ago a Swiss cruiser installed solar panels, batteries and wiring so that he has lighting for his restaurant. He has a generator to run his freezer and refrigerator. As the president of the rural business association he flies to the capital, Port Vila, every week. Safari Tours signs up tourists in Port Vila to stay in his guesthouse. Most of the guests at his restaurant come from the boats anchored out a short distance from the beach. His wife is a very accomplished cook.

The next day it was pouring rain, but we managed to row ashore in our foul weather gear to make reservations for dinner for us and friends who had just arrived. The next evening we had a delicious dinner of wahoo served two ways, plus spaghetti, taro chips, sweet potato patties, cucumber slices and watermelon, all for about $6 each. Port Vila has several expensive and fancy restaurants, but it would be hard to match the delicious dinner prepared by Tasso's wife. The menu seldom varied, but it was always wonderful. During our week at Lamen Bay we ate there twice more with different groups of cruisers. The third time we were joined by guests at the resort, a Cathay Pacific Airlines pilot from Hong Kong along with his wife and two children. He told us they were shocked when they arrived and saw 14 boats at anchor at what was supposed to be a secluded beach bungalow, but soon found that they enjoyed the company of the cruisers, sharing snorkeling and meals with us. Tasso told us about living on the small island of Epi. Five dialects are spoken on the island, so they communicate in Bislama. The residents make their living from growing kava, which is exported to America, Japan and China; copra, which goes to Europe; and trocha shells, which go to Korea to make buttons. The villagers all have gardens where they grow fresh food, some of which they sell or trade to visiting sailors.

The snorkeling at Lamen Bay is excellent, with lots of colorful tropical fish, and a reputation of being free of sharks, unlike some of the anchorages in Vanuatu where sharks abound. Lamen Bay also has a resident dugong, an endangered species of sea mammal, similar to a manatee, that lives on sea grass. The locals call it a sea cow. When a dugong made its appearance, we all swam over to watch it as it plowed the sand, devouring the seaweed like a vacuum cleaner, leaving a series of long swaths of bare sand in its path. It was about 10 feet long and weighed about 800 lb. It was apparently undisturbed by all the snorkelers who hovered above it. It just continued to avidly graze on its vegetarian diet. It was fascinating to watch this creature up close.

The day before we left I dug out a large, unused guest book that we were given by friends years ago, but found to be too large to fit on the bookshelf on our boat. At many of the popular anchorages we had visited, the chief of the village has a guest book that visiting cruisers fill out, writing comments on their visit and sometimes including a picture of themselves or their boat. We noticed that Tasso didn't have a guest book, and thought he would enjoy having a record of his many guests at his small resort. We renamed it the Lamen Bay Guest Book, and covered it with clear plastic to protect it from the elements. That evening we were invited for drinks along with a dozen other cruisers on one of the larger boats anchored in the bay. We brought along the guest book, and passed it around to the cruisers to fill out. That day we also learned that the American Peace Corps teacher at the Lamen Bay middle school was looking for games to entertain her students. We had a small game of Othello that we dug out of storage to give her. The next morning we went ashore with the guest book and the game. Tasso was delighted with the guest book. We took the Othello game to the small school, which was about a 20-minute walk inland from the bay. We could hear singing coming from two of the classrooms as we approached. The American teacher wasn't there, but we talked to the Australian teacher. We learned that this was a boarding school, with 150 students who come from Epi and nearby Paama, Ambrym, and Efate islands. It seemed like a pleasant place of learning, with singing and games.

It was time to leave Lamen Bay and work our way further north. We had had a great week at this anchorage with interesting visits with Tasso, wonderful meals at his restaurant, and a chance to learn a little bit about life in this small village. We also enjoyed some spectacular snorkeling, and the camaraderie of the many cruisers who shared the anchorage with us. This was one our favorite anchorages in Vanuatu.

## Visiting the Banks Islands

On our second trip to Vanuatu in 2000 we wanted to visit Banks Islands which we had not had time to visit on our trip two years previously. Cruising friends who had been there told us visiting the islands was a must. Every year hundreds of boats visit the major towns of Port Vila and Luganville, but only a few sail north to the remote Banks Islands. We found out from the Cultural Center in Port Vila that they were holding the first ever Torba Arts Festival in September. Torba is the province that contains the Banks and the Torres islands. We and several of our cruising friends planned to attend.

Our 10-day sail from Coffs Harbor, Australia to Vanuatu had been fairly easy. After checking in at Port Vila we spent July and August visiting the islands of Emae, Epi, Ambrym and Malekula, ending up at the town of Luganville on the south end of Espiritu Santo. From Luganville we had a two-day sail to Vanua Lava, the largest of the Banks Islands. We arrived the day before the festival began, and watched the harbor at Sola gradually fill up until there were 19 boats at anchor. Every day we cruisers went ashore in our dinghies to the festival grounds, a large plateau surrounded by trees and hills. The opening ceremony involved the sacrificing of a pig. A large pig is tied up and dispatched with a heavy club by one of the chiefs. We cringed at the cruelty of this ceremony, as the pig had to be hit repeatedly, with much squealing, before it died. Dancers from five different Banks Islands performed their traditional dances every day in costumes made of local plant materials. They wore grass skirts, leaf necklaces, and elaborate headdresses some in the shape of tuna, sailfish, sharks, or sea urchins. The chiefs from each village joined in the dancing. They could be identified as chiefs by the large cycad palm fronds they carried. The dancing was vigorous with much stomping of bare feet, and the rattle of strings of nuts wrapped around the dancer's ankles, accompanied by large wooden drums that were beaten by men on the sidelines. Our favorite was the snake dance, in which all the dancers had painted horizontal white stripes all over their bodies. With white stripes against their black skin they resembled sea snakes we had often seen in tropical waters. In the heat of the tropical sun the dancers must have been sweltering, but they danced for several hours every day.

Interspersed with the dancing the islanders shared other facets of their culture. One morning they showed us how they prepare laplap, the national dish. We watched the men grind up the cooked taro in a huge wooden bowl with a heavy pestle the size of a baseball bat. We couldn't help notice that as they worked the sweat from their chest was dripping into the bowl of taro. It was a bit off-putting, but was probably harmless as all of the food is wrapped up in leaves and cooked for hours in an underground oven over hot rocks. The taro was topped with ground nuts, fish, or spinach before it was cooked. When they opened up the oven, they removed the long cylindrical bundles of laplap, and sliced it with wooden laplap knives. They served it to us to sample. I found it quite tasty, although many cruisers thought it was too bland.

Several of the men were expert carvers. We watched in fascination as these carvers fashioned intricate fish, bowls, and laplap knives with their primitive tools. They sold their carvings to us at very reasonable prices. The men demonstrated basket weaving, using vines, and fire making by rapidly twirling a pointed stick in an indentation of a block of wood. The last day the villagers from nearby Mota Lava demonstrated catching fish by casting a net into the shallow water by the beach. After pulling the net partially closed with vines along the upper edge, everyone waded into the net to kill the fish with sticks, clubs or bows and arrows. By the end of the week we felt we had learned quite a bit about the lives and customs of the Banks Islanders, and the inhabitants of the different islands learned about customs of their neighbors.

After six days of calm weather we had a change to north winds, making the anchorage rolly and untenable. We left to sail around the top of the island to the other side to anchor at Waterfall Bay. It is well protected from most winds. We had heard on the cruiser's net that Don, an engineer from the sailboat *Klondike*, was installing a waterfall generator at the base of the waterfall. One of the villagers had read a magazine article about waterfall generators the year before. He had asked Don about the possibility of installing one at Waterfall Bay. This year Don and Katy had brought a used car alternator, a discarded fire hose from a fire station, and a large used battery on their sailboat. The next morning we followed the group of cruisers who climbed the hill to the top of the waterfall carrying the heavy fire hose. Don and a couple of other men placed one end of the hose underwater in the stream with a filter over the opening, and teetered over the edge of the waterfall to drop the heavy hose down to the ground, where small paddles attached to an alternator were located. When the nozzle of the hose was directed so that the stream of water hit the paddles, the alternator began producing electricity which fed into the battery. The villagers were ecstatic at the idea that they could have a continuous source of free power and light in their village. To celebrate, the villagers invited everyone to a big feast ashore honoring Don and Katy, who had made it happen. They served us sweet potato laplap, curried shrimp, sautéed fish, rice and cucumber slices. We spent several more days at Waterfall Bay, topping up our water supply from a wonderful spring that came down to the bay at water level, only accessible at low tide. We visited with the other cruisers and villagers, swam daily in the pool of clear water at the base of the waterfall, and used the fresh, clear water to wash our clothes as it passed over the rocks near the beach to the sea.

Our next stop was Vureas Bay, just ten miles south of Waterfall Bay. For a couple of days we shared the anchorage with Gordon and Miriam, while the other cruisers headed south back to Luganville. As soon as we had anchored Godfrey, the Chief of Vureas Bay, came up to our boat in his dugout canoe and invited us ashore. We rowed ashore and met Godfrey's wife, Veronica, and had a tour of their home and garden. They had built three guest bungalows near the black sand beach, and had planted a beautiful garden of rows of hibiscus, mango, banana, papaya, and nut trees. We learned that there was a village of about 700 people up the hill from Godfrey and Veronica's home, but we decided against visiting it because of the slippery muddy trail after a rain. When we did reach it the next day, after a 45-minute hike up the rocky trail, we found a small school, a store, and a church under construction using cinder blocks. The homes were close together, and were mostly made of woven palm fronds, with dirt floors. We were welcomed by several of Godfrey's sons and their families who lived in the village. Dan showed Godfrey's son Brian how to sharpen his chisel with a whetstone. Over the next couple of days we traded drill bits, fishhooks, chisels, T-shirts, balloons, and pencils for fresh vegetables, fruit, and freshwater prawns with the villagers.

After Gordon and Miriam left Vanua Lava, our boat was the only one at anchor. Godfrey asked for a showerhead for his guesthouse, and we had an extra Sun Shower we gave him for the purpose. He had a flashlight that needed repairing, which Dan was able to do. Godfrey was grateful for any help, and offered to adopt Dan as his son. Dan was taken aback, and though he felt honored, he pointed out that he couldn't be his son as, at 64, he was at least 15 years older than Godfrey. Instead, Godfrey invited us for a farewell feast the next evening, as we were departing after that to head south. It was a memorable dinner with Godfrey and Veronica, their sons Brian, John and Stephen. They had decorated their dining room table with strings of hibiscus flowers and leaves from their garden. Godfrey made a speech in Bislama, translated into English by Stephen, about how grateful he was for our visit, supplying him with a shower and other help. They served us a delicious dinner of fish in onion sauce, rice, taro, wild yams, bananas and taro laplap topped with grated coconut. We had a nice visit afterwards before we had to return to our boat. We had enjoyed four days at Vureas Bay and had felt almost part of Chief Godfrey's family.

Visiting the Banks Islands was certainly the highlight of our cruising that year. Three years later we went back to the Banks, this time attending a festival in Waterfall Bay. After the successful first Torba Arts Festival the islanders in Vureas Bay and Waterfall Bay decided to hold a festival every year. Each festival was different, but all were eagerly attended by cruisers from around the world. It was a great way of sharing their culture and entertaining their visitors. The waterfall generator ultimately failed due to a land dispute. Two families each owned half of the waterfall, and couldn't agree to share the power, so, unfortunately, it was abandoned. They are still in darkness at night. Godfrey learned that cruisers who were older than he didn't wish to be adopted as sons, so the following year when our friends Gordon and Miriam visited Godfrey he adopted them as his parents. When they returned every year they were greeted as Mami and Papi, and there were tearful farewells when they left. Gordon and Miriam learned that being adopted by locals carried with it certain obligations, and every year they brought dishes, linens, silverware, sheets and whatever else Godfrey and Veronica requested for their guest bungalows. They were glad to help out, and Godfrey and Veronica were truly grateful. All the cruisers we knew loved the Banks Islands and the people who lived there and returned repeatedly. The Banks were our favorite islands in the South Pacific.

## Sailing to New Caledonia

After we returned to Coffs Harbor from our boomerang trip in June 2003 we had a great deal of work to do before we could depart again. First I had to soak off the dried blood from my head injury. I stood in the shower for 15 or 20 minutes with warm water coursing over my head, gently rubbing my hair and scalp with my fingers until the blood dissolved. Sue, a nurse on a sailboat in the marina at Coffs Harbor, advised this method so she could assess the extent of the injury. She looked at it again after it was clean and said it was healing well and was not a deep gash. That was a big relief, and with the dried blood gone I no longer looked battle-scarred.

Next we had to replace the leaking seacock under the sink. We scheduled a haulout of our boat on the marine railway adjacent to Coffs Harbor marina. With one day's work Dan was able to remove the faulty seacock and replace it with the brand new, improved model we had stored on the boat. The only remaining problem was our batteries, which we had noticed, weren't holding their charge during our boomerang trip. While at a marina, or coastal cruising we use the batteries only for reading lights at night and the ham radio. Our solar panels keep the batteries charged. During a passage the batteries are much more heavily used. We turn on our masthead navigation lights from sunset to sunrise, and we turn on the radar when we see a ship so that we can avoid a collision. We have the VHF radio and GPS on day and night, and we use the ham radio for the net, for weather reports and to send and receive email. The batteries have to be in good shape during a passage so that the solar panels can keep them charged up. We decided to do capacity tests on these almost new batteries. After several overnight tests, leaving all our navigation lights on to provide a constant five-amp drain, we determined that one battery had only 60% and the other 40% of its rated capacity. When we complained to the battery company in Sydney where we bought the batteries they asked us to send them back so they could test them. They replaced one with a new one, but claimed the other one passed their test. We ran another capacity test on both of these batteries, but they were still well below their rated capacity. The battery company finally agreed to take them back and refund our money. We immediately ordered two new ones from a marine store in Brisbane, tested them and found they were close to their rated capacity. The serial numbers indicated they were much newer than the first ones we were sold. We figured that the bad batteries we had purchased in Sydney had been sitting on the shelf for months if not years, and had sulfated because they hadn't been kept charged up. They weren't usable for our purposes.

Our battery hassle had lasted about four weeks. During that time two boats left for New Caledonia after weeks of preparation. A steel boat, with three surfing buddies who had never sailed before, left and returned within 24 hours when their boat had electrical problems. Our friends Frank and Therese from Scotland left and sailed for two days when they made an accidental jibe in strong winds and ripped their mainsail. They had to return and take the sail to the sail maker's for repair. At one point the harbormaster said more boats were returning than leaving. We knew how they felt—it is so discouraging to spend weeks getting ready, finally depart, and come limping back into the marina within a few days. The boomerang club at the marina was growing.

Finally the winds were predicted to be in the right direction, with no storms in the offing. We re-provisioned and departed for Noumea, New Caledonia again in late July. This time we had no gales, no injuries, and no gear failures but it was a very rough passage, nevertheless. We had multiple days of squalls, waves occasionally breaking into the cockpit, a scary lightning display, and some rain. It was nothing like the pleasant downwind sails we had had in the trade winds between Mexico and French Polynesia. Just before we left Coffs Harbor we had gone to happy hour on a friend's boat along with two other couples. While our husbands discussed engines and navigation we women got together to discuss more interesting things. All four of us agreed in our assessment of cruising—we hated the passages, but loved the cruising life. This passage only confirmed my opinion. Even Dan admitted he was not enjoying the passage. We talked about making this passage our last one.

After days of being tossed around in big seas and strong winds, in the late morning of the eighth day we reached our waypoint just outside the narrow pass in the fringing reef that surrounds New Caledonia. Because we'd had a serious problem of the engine overheating whenever we tried to motor in rough seas, we decided to sail through the pass until we got into the calm waters of the lagoon. As a precaution I put three waypoints into the GPS to guide us through the winding passage and into protected waters. Dan steered for the Amadee lighthouse with the mainsail well out, heading downwind. Before reaching the lighthouse we had to turn north to pass between coral heads marked with beacons and continue on a 12-mile winding course within the lagoon to the anchorage. As we sailed through the pass toward the lighthouse I looked at the course to steer on the GPS, and noticed we were about five degrees off course, and when we turned towards the pass between the coral heads we were again way off course. I kept saying: "We're off course, steer five degrees to the right," but Dan refused to listen, saying: "I'm heading in the right direction. I don't need your instructions." I was really annoyed. After I'd taken the trouble to enter the waypoints we would need to get us safely into the lagoon, Dan ignored all of my directions. I gave up. I went down below and said, "You're on your own—I'm not helping you navigate anymore." Then I glanced at the chart on the chart table, and was mortified to read: "These latitudes and longitudes have to be corrected by .15 minutes to correspond to the chart datum in the GPS." It had been three years since we last visited New Caledonia and I had forgotten that the French use different chart datum from everyone else in the world. My waypoints were off by 500 to 600 feet. I felt really foolish for making such a fuss about Dan not following the GPS waypoints. If he had listened to me we most likely would have ended up on the reef. Dan's ability to steer using eyeball navigation instead of the GPS saved the day.

Now that we were in the lagoon, which was like a big lake, we turned on the engine to motor the last 12 miles to the anchorage to make sure we would be anchored before sunset. We were moving smoothly through water in the wide lagoon when we heard a weird grinding noise and a series of loud thumps coming from underneath the boat. I had a sickening feeling that we had hit a reef, but when I checked the depth sounder it read 78 feet, so that was impossible. Dan thought we had hit some floating object, but we looked behind the boat and saw nothing. We turned off the engine, and I took the helm to sail while Dan checked the engine. He discovered that a plastic bag containing four large spools of string that he had hung in the engine compartment had fallen from a hook onto the propeller shaft coupling. The string from one of the spools had hooked on a bolt, causing the whole spool to unravel, and tangle with hundreds of loops around the rapidly turning propeller shaft. The other three spools had bounced repeatedly on the shaft causing the loud thumps. Armed with his pocket knife and a large plastic salad bowl Dan was able, in about 10 minutes, to cut the string off in segments and fill the bowl with twisting strands of white twine, which he brought up to the cockpit to show me. We laughed and laughed at the result—it resembled a large bowl of spaghetti. It looked good enough to eat if we poured spaghetti sauce over it. Now we could start the engine again and continue on to Orphelina Bay. We dropped our anchor just before sunset in the well-protected anchorage. We celebrated our arrival with a glass of wine, very happy to have survived our entry into Noumea.

Even with the most careful preparations we always seemed to encounter unexpected problems on a passage. We were greatly relieved we had avoided wrecking on the reef, all because Dan had the good sense to ignore my faulty waypoints. And now we had a bizarre souvenir of our arrival in Noumea to show to our friends—a large bowl filled with strands of white string resembling spaghetti.

## Shipwreck on Tanna Island

After our boomerang trip in 2003 we departed from Coffs Harbor for New Caledonia in late July. We stayed at the marina in Noumea for a week, and then sailed to the Bay of Prony on the east side of Grand Terre. From there it is only 250 miles to island of Tanna, near the south end of Vanuatu. We had been told that we could check in to Vanuatu at Tanna, rather than Port Vila, the usual port of entry.

In the early morning of the second day of sailing we approached the island of Tanna. As we sailed up the east coat of the island towards Port Resolution, we worried that our gps position didn't fit with our chart. Although we planned to stay at least five miles off, we appeared to be much closer to the island than indicated on the chart. Then I checked the date on our chart—1885! In those days sailors were not able to determine longitude accurately, so we could not trust our chart. As we got closer to Port Resolution we were shocked to see the wreck of a large sailboat on the rocky shore just south of where we thought the entrance was. The shiny black hull with mast and rigging intact was lying on its side, with gentle waves breaking over it. It was scary to try to enter a harbor with an ancient, inaccurate chart, especially after seeing a wrecked sailboat near the entrance. I called on the VHF asking if any boat in the anchorage had waypoints for entering the bay. Mary, on *Tranquility*, responded, reading off the three waypoints they had been given. I entered them in our gps, and we found it easy to follow the waypoints into the anchorage.

When we arrived there were 22 sailboats in the anchorage of Port Resolution, many from the U.S, and others from Australia, Canada, England, Ireland, and France. Most of the cruisers came to visit the active volcano of Yasur, but we learned that the annual Toka dance festival was happening in a few days, which was another big attraction.

Every cruiser knew the story of the wreck of the *Coker Lady*. The story was second hand, told by the villagers who were there at the time of the disaster, so it may not be accurate. The *Coker Lady* was a 58-foot fiberglass sailboat sailed by an inexperienced British couple and two Fijian crewmembers. They were anchored in Port Resolution a month before our visit, when a shift in the winds brought strong northerly winds. The anchorage is open to the north, and soon waves from the north built up, making the anchorage untenable. All of the boats left the anchorage except *Coker Lady*. The crew decided to wait it out, even though the pitching and rolling of the boat was extremely uncomfortable. The villagers suggested to the crew that they come ashore and stay in guesthouses until the storm was over, but they refused the offer. However, after a few days of violent motion they made the decision to leave in the middle of the night. As they went out the entrance to the bay the boat was hit by a huge wave, which broke over the boat, knocked it on its side, and destroyed their electronics. They were disoriented, and thought they were turning right to leave the bay, but instead headed directly into the east shore of the island. The boat went aground on a rocky shelf. They were able to climb out of the boat onto land and eventually make their way in the dark up to the village. They were grateful to be alive, but the boat was destroyed. Fortunately it was insured, so it was not a financial disaster. It certainly destroyed their cruising dreams. The villagers housed them ashore, and in a few days took them to the airport at Lenakel, where they flew to Fiji, and home to England.

The boat was a write-off, but the islanders had salvage rights. They acted quickly to remove anything that was valuable. It was an amazing feat to remove the engine and carry it through the jungle and up the hill to the village, where it was stored on the ground under a tarp. They removed the diesel generator, electronic equipment, winches, shackles, turnbuckles, and other fittings that were salvageable. The smaller items were displayed for sale on the tables near the yacht club, where the women sold produce to cruisers. Earlier a cruiser had put prices on all the items, at a discount price based on the West Marine catalog, but it was difficult to sell most of the items. They would only fit a boat of comparable size. Most cruising boats are fully equipped and don't have room for extra equipment. The heavy engine and generator would be particularly hard to sell.

The day after we arrived we walked down the now well-worn path to the rocky beach see the wreck up close. The boat was essentially an empty shell. The fiberglass hull was slowly breaking apart by the constant wave action. It was sad to think of the shattered dreams of these cruisers, and the frightening end to their cruise. It was a lesson to everyone to be cautious in stormy weather, and not make bad decisions in the middle of the night.

A week after we arrived I went down to the beach again to see the wreck with our newly arrived friends, Frank and Therese, from *Dansa Na Mara*. The boat was continuing to break up with the constant pounding of the waves. On the way back we stopped in the village and watched two men from cruising boats working on the generator.

They hooked up a water supply for cooling, and poured some diesel into the tank, while a dozen or so men and boys from the village sat around the edge, watching every move. Eventually the cruisers were able to get the generator running. Someone brought an electric toaster and some bread, and with great ceremony the cruisers put in two slices, and a few minutes later cut the toast in small pieces to pass out to all the bystanders. It was a triumph for the engineers/cruisers.

A couple of days later the hard task began of moving the generator into the schoolhouse to protect it from the elements. The villagers had to get special dispensation from the chief to work on Saturday, as they were Seventh Day Adventists. They began at noon, after much discussion as to how to move it. They had 15 strong men for the task. They carried a long pole from which the generator was suspended by ropes. They had several long poles underneath the generator, with strong men on both ends. They marched along, chanting as they went, and managed the long trek to the schoolhouse, where the generator would stay for the rainy season.

A few days later we went to the village to see the much larger engine from *Coker Lady*, a 150 h.p. Yanmar that was worth at least $12,000. It would be a huge sum of money for the villagers if they were able to sell it. We saw five ni-Vans taking a ride around the bay in the large dinghy with an outboard motor. We were glad to see the villagers enjoying the one useful item from the tragic wreck of *Coker Lady*.

## Adventures on Tanna Island

On our first visit ashore on Tanna we took the short walk up from the beach to the small village where there are a few small bamboo homes, and tables set up to display produce from the villager's gardens. Chief Ronnie greeted us and told us about the school which he owns. Originally the church supported it, but they discontinued their support because the villagers were not Christian. Ronnie hired teachers for the school with money he received from tourists, who paid to see the dugong. (Tourists paid $7.50 to see the dugong at Port Resolution whether or not the dugong appeared). Now the dugong is dead, so Chief Ronnie turned the school over to the government to run.

Shortly after we arrived one of the cruisers arranged to have the villagers prepare a feast at the yacht club for the cruisers. There was a charge of about $10 per person. The Yacht club is an impressive open air structure with bamboo walls about three feet tall, a corrugated tin roof, and yacht club burgees from all over the world hanging from the ceiling. It's on a ridge overlooking Resolution Bay. There were several long tables at the yacht club, and many wooden chairs, some of them rickety. We talked to Graham and Iris on *Timewise* and learned that he was going ashore in the morning with another cruiser to try to repair the rickety chairs. Dan joined them, and the three men spent an hour screwing plywood triangles to the frame of the chairs to strengthen the leg join.

The day of the feast everyone gathered plates, cutlery, water, and glasses from their boats and rowed over to the beach. A group of villagers met us there, and led us up the path from the beach to the yacht club, singing enthusiastically. They were accompanied by four young men playing guitars. Before the meal Chief Ronnie gave a welcoming speech, saying they have nothing to export, so they're glad to have us come ashore to enjoy their food. The food was all local, and was beautifully displayed on large banana leaves on long tables covered with tablecloths. The women had prepared curried chicken, roast pork ribs, broccoli, green beans, taro, lettuce and tomato salad, and cabbage and carrot slaw, with papaya and pomplemousse slices for dessert. The cruisers piled the delicious offerings on their plates, and sat in the repaired chairs along the edges of the room. While we ate singers and guitar players entertained us. In addition the chief and another man from the village, in bare feet and dressed in grass skirts and ankle rattles, performed a short dance to the music. A crowd of women and children stood outside the building watching it all. The villagers at Port Resolution certainly knew how to entertain. It was a great way for the cruisers who had come from all over the world to get to know each other.

This year Tanna would be an official port of entry. We could check into customs, immigration and quarantine on the island, rather than Port Vila, which was the usual port of entry for cruisers. However we would have to travel to Lenakel, on the opposite side of the island from Port Resolution. The chief arranged for a truck to pick us up at 7:30 on Monday morning, at a cost of about $30 each. Ten cruisers piled into the back of a red pick-up truck. The chief's son, Stanley, drove us on a 90 minute bumpy ride across the island. The scenery was beautiful, as we passed through a jungle with large banyan trees, then up the slope and across a black sandy lava field, on the unpaved roads. To keep from sliding around on the wooden benches in the truck bed we held on to metal rails mounted a few feet above the bed of the truck. It turned out that the customs and quarantine offices were closed, but we were able to check in with immigration. Lenakel had a small bank, where we changed our Australian dollars into vatu, and a butcher, baker and produce market, where we purchased food not available in the small village at Port Resolution, so the trip was not wasted.

Our next journey was two days later for the Toka dance festival. This is part of a three-day annual festival on Tanna, known as the Nakiwar. This is one of the most important events on Tanna, held every year, usually in August, but not announced until a few days before. We were fortunate to have arrived in time to attend the Toka dance. We signed up for the ride to the site and paid $70 for the transportation and the entry fee for two of us. The truck arrived at 6:30 a.m. to pick up the cruisers who had gathered at the beach. Fifteen of us crammed ourselves into the back of the pick-up truck. This time Dan and I brought our cockpit cushions to make the ride more comfortable. After a short distance up a steep slope the truck stalled—the engine was not powerful enough to carry that much weight. Six of the men climbed out and began walking up the hill. When the truck came to the level part of the road the driver stopped and picked them up. By the time we arrived the dance was in full swing. We watched at a distance as five groups of male dancers from different villages danced continuously. The men sang and danced barefoot, in grass skirts and feather headdresses. The women on the sidelines wore grass skirts and had brightly painted faces. We watched for hours, while the vigorous stomping by the men stirred up a huge cloud of dust and fine sand. We had to constantly move to the upwind side of the dancers to keep from choking on the clouds of dust. The dancing began at dawn and would continue on to the next morning. It was a spectacular sight. We left before noon, for the 45 minute ride back to Port Resolution. On a walk through the jungle a few days later we talked to one if the villagers who had danced the last dance of the Toka festival. He told us the dancers practice weekly for a year before they're ready to perform.

The main attraction on Tanna for most people is the active volcano, Yasur. Many of our friends paid for the truck ride and a hefty entry fee to visit the volcano. They told us it was awesome. While standing on the rim of the red glowing crater they watched molten lava shooting up into the air every few minutes, like a fireworks display.

Before we scheduled a trip to the volcano we read our South Pacific Handbook and Lonely Planet book and learned that visitors are driven up the lava sand slope to a trail, where it's a short walk up to the rim of the crater. The crater is about 1000 feet across and 300 feet deep. There is no guard rail around the crater. It's best to go at night when the fireworks display is more spectacular and you can see the glowing rocks that are thrown up and have a chance to avoid them. In 1995 a Japanese tourist and two guides were killed when hit by a huge chunk of lava while standing on the rim of the crater. The sulfur fumes can be choking and the noise deafening.

When Dan and I visited Yellowstone National Park seven years before, we found out that Dan has a serious allergic reaction to sulfurous fumes. One morning we drove to a large area to walk on boardwalks among the gaseous, bubbling fumaroles. As we approached Dan began to choke, and his eyes became red, itchy, and watery. We didn't even get out of the car, but turned around to escape from the fumes. The possibility of being hit by molten lava was frightening enough, but having Dan choke on sulfur fumes in this remote area was the deciding factor. Tanna had already provided us with plenty of adventure and entertainment. We decided to forego visiting the volcano. Later we could look at the spectacular displays in photos or videos without risking our lives.

## Dangers of Cruising

Besides the illnesses and tropical diseases cruisers are subject to, accidents are a real hazard of cruising. It is rare for sailors to be injured on a passage, especially in good weather and smooth seas. In the Queen's Birthday storm off New Zealand in June 1994 a fleet of sailboats were caught in cyclonic winds and huge seas. Several people suffered broken ribs, broken arms and injured backs. We watched a film of the heroic rescues of sailors by the New Zealand Search and Rescue teams who lifted sailors off their boats to safety by helicopter.

One evening when we were in Port Vila, Vanuatu, we heard a young woman calling for help to guide her sailboat into the harbor. She and her husband had departed that morning for their trip back home to New Zealand. In the rough seas her husband had fallen out of his berth and was in great pain. He had had a hip replacement, and thought he had broken his pelvic bone. He was X-rayed at the hospital in Port Vila, and it was determined that he was badly bruised, but fortunately had no broken bones. When our boat was knocked down by a rogue wave off the New Zealand coast I hit my knee on the furniture. Off the Australian coast, I fell off my berth, hitting my head on the corner of the computer shelf. Both times we were in very rough seas and gale force winds. My injuries were minor.

One of the biggest hazards on board a boat is a wind generator. These are propellers mounted on small generators. Wind turns the long blades, producing a current to recharge the boat's batteries. Usually they are mounted high off the stern of the boat out of the way of the sails or rigging. When we were in Cabo San Lucas we saw a sailboat with a family of five children aboard that had a wind generator mounted on the foredeck only three or four feet above deck level. The children knew of the danger, but when one of their friends visited the boat she ran forward under the wind generator and gashed the top of her head on the spinning blades. She had to be rushed to the clinic in Cabo San Lucas for stitches. A skipper of a sailboat with the wind generator mounted on the stern reached up to adjust the angle of the generator. He nearly severed a finger in the fast-revolving blades. A couple in Tahiti were standing on the stern of their boat to say goodbye to their friends as they departed from the harbor. The skipper raised his arm to wave farewell. His fingers were caught in the revolving blades and were severely mangled. This of course delayed their departure until his hand healed. When we purchased our wind generator years later I had a vivid memory of these sailors' stories. I insisted we mount our wind generator in the stern of the boat about eight feet off the deck where neither of us could ever be injured by the revolving blades.

Another accident occurred to a friend of ours when he motored ashore in his inflatable dinghy in Vanuatu. There was surf breaking on the beach. As he neared the beach his inflatable was caught by a wave that turned the dinghy sideways, then flipped it over on top of him. His leg was slashed by the blades of the outboard motor. Luckily he was able to get medical treatment at a clinic in a small village nearby. Because our dinghy is fiberglass it can be propelled either by rowing or outboard motor. Whenever we have to land on a surf beach we leave the outboard off and row ashore. We have been dumped by breaking waves while rowing the dinghy, but have never been injured.

Another hazard that is rare is lightning. We know one couple whose boat was hit by lightning twice within a few weeks in New Caledonia. If lightning strikes the top of the mast it probably won't injure inhabitants, but often destroys all the electronics in the boat, such as radar, radios, GPS, and computers. It can cost a great deal to replace them. During a passage it is frightening to see lightning striking all around when you're in the only boat for miles. On one passage to New Caledonia we saw a spectacular display of lightning lasting for an hour. Fortunately it was several miles away. One summer season when we were in a marina in Brisbane we experienced lightning storms nearly every afternoon. Heat built up all afternoon, then dark clouds filled the sky, and the heavens let loose with bright flashes and loud claps of thunder, sometimes accompanied by hailstones. We cowered inside the boat, counting the seconds between the flash and the thunder. One bright flash of light and a simultaneous deafening clap of thunder appeared to be a direct hit to the marina. The next day we learned that it had hit the mast of a boat in the next pier over from us, less than 100 feet away.

Most injuries occur not when on a passage, but when the boat is in the boatyard for the annual maintenance—sanding and antifouling the hull below the waterline, and cleaning and polishing the topsides above the waterline. Usually, there is crude scaffolding alongside the boat, consisting of long planks supported by overturned barrels. We've known of several sailors who've fallen off the scaffolding, with serious head injuries or broken bones.

Creatures that lurk in the water around the boat are scarier than anything on the boat. The most horrific encounter with a sea creature, which we heard about on the radio, happened to a Swiss sailor who anchored in a bay in Papua-New Guinea. He dove down in murky water to check whether his anchor had set, and was immediately grabbed in the jaws of a crocodile. His wife watched in horror as the crocodile dragged his body ashore to be eaten later. He had ignored the standard advice to stay out of the murky water in areas where crocodiles live. Another horrible story we heard on the radio was of a German cruiser who had cruised for years. He regularly speared fish for his dinner, and dragged them behind him in a bag. While he was snorkeling in a remote anchorage he was attacked by a shark, which bit a large chunk of flesh from his arm in spite of his efforts to fend it off. He was saved from bleeding to death by a cruiser in a boat anchored near by. Fortunately, the rescuer was a trained Emergency Medical Technician, and was able to stop the bleeding and treat the German's gaping wound. The cruiser had violated the standard rule that as soon as a fish is speared, both hunter and fish must get out of the water, because sharks are attracted by the thrashing fish and the blood.

In Mexico we enjoyed wading in the shallow lagoon near one of the anchorages in the Sea of Cortez. A friend of ours stepped on a stingray while wading, and experienced excruciating pain in her foot. When she and her partner returned to the boat they read that the best treatment was to soak the foot in water as hot as she could stand. It was like a miracle cure—the hot water apparently denatured the protein of the toxin, and the pain disappeared immediately. We learned that it is best to shuffle our feet when wading in the water to avoid accidentally stepping on stingrays.

A dugong is a large marine mammal that we were delighted to see in Lamen Bay in Vanuatu. We and other cruisers snorkeled at the surface above the dugong and watched it comb the sea floor with its large mouth, sucking up the sea grass. Later we learned that a friend, Don, on the boat *Sour Dough*, had a frightening encounter with this same dugong. The dugong seemed to have amorous intentions towards Don. The dugong grasped him with his flippers and pulled him down about 10 feet to the bottom of the bay. He held him underwater long enough that Don ran out of air, and thought he was going to drown. With a terrible struggle Don was able to get out of the dugong's grasp and pop up to the surface to breathe.

Another bizarre accident happened to a fellow cruiser we had met in Mexico. He caught a large fish, which he hauled out on the side deck. His foot brushed against the sharp points of the fish's dorsal fin, causing a small puncture wound. His foot began to swell and redden and became extremely painful. When he spiked a high fever he and his girlfriend called for help on the emergency channel. He was airlifted by navy helicopter to a hospital in Hawaii, where it was determined he had been poisoned by a rare toxin in fish. The bacteria cause an infection that can be fatal if not treated. He received the necessary treatment, and the navy dropped two crewmembers off onto his boat to help his girlfriend sail the boat on to Hawaii.

When we were in the northern Sea of Cortez we went to a nearby bay to snorkel amongst the resident sea lions. I was a bit uneasy when a sea lion approached me in an aggressive manner, but I didn't know there was any danger. Since then I've learned that sea lions can be aggressive, and in California have been known to bite swimmers who have encroached on their territory.

Fortunately, we didn't encounter any crocodiles except in the zoo, and were wary enough of sharks that we never had a problem even when we snorkeled on a reef teeming with them. Although Dan pulled in a number of fish which he filleted on the side deck, he never had a scratch or wound from a fish fin.

In spite of all the things that can happen, cruising is probably one of the safest activities we could have chosen. It is certainly safer than rock climbing or skydiving. However, if we had thought about all the possible hazards of cruising we probably would have been afraid to ever leave home.

## Helping the Banks Islanders

In 2003 and 2005 we returned to the Banks Islands in Vanuatu to visit the islanders we had so enjoyed on our first visit in 2000. Vanuatu is about 1,000 miles north east of Australia, and consists of a chain of a dozen volcanic islands running north and south. The Banks Islands are the three most northern islands of Vanuatu (except for the small Torres Islands). They are isolated from the rest of Vanuatu by 50 miles of ocean.

The centers of population, Port Vila and Luganville, are bustling towns with cars, trucks, buses, grocery stores, hardware stores, restaurants, resorts, and Internet cafes. By contrast the Banks Islands are primitive. There are no roads, cars, buses, phones, electricity, doctors, hospitals, or libraries. People live in small villages near the shore, and grow their own vegetables, usually on the steep slopes away from the shore and a long walk from home. They grow taro, manioc, yams, island spinach, and long beans and kava. They harvest coconuts for copra, and pick pamplemousse, papayas, and nuts from trees. They fish from their outrigger canoes. Their homes are mostly thatch-roofed bungalows, bare of furniture except for a dining table and chairs. The women cook on an open fire or an underground oven, which is a hole in the ground with a layer of hot rocks, food wrapped in taro leaves, and the whole covered with taro leaves. The people were the friendliest people we had met anywhere. When we anchored in a bay near a village we were usually greeted by villagers in their outrigger canoes. They offered us fresh produce from their gardens, and we asked what they would like in trade. It was always gratifying to trade with them because they were so appreciative of whatever we gave them. They often invited us ashore for a meal or a walk through the village.

Over the years many cruisers from Australia and other countries have visited Vanuatu. Cruising sailors begin arriving in June or July and leave by October to avoid the hurricane season in Vanuatu. On their visits they've learned about the life and customs of the people, and what they need in the way of basic necessities. Before sailing to the islands we stocked up on items such as T-shirts, old magazines, fishhooks, vegetable seeds, soap, matches, rice, canned fish, school exercise books, pens, and pencils to trade for fresh food. It is a mutually beneficial arrangement, providing us with fresh fruit and vegetables in exchange for items the natives are unable to obtain in these remote islands.

In addition to cruisers helping the islanders, every winter a medical team of doctor, nurse and assistant are sent to the Banks Islands by the Seventh Day Adventists in Australia. While we were there the three-man medical team was transported by Bob and Judy on the trimaran *Siddiqi*. They traveled to villages on different islands and set up one-day clinics to treat various problems—minor wounds, head lice, and infections. The assistant, who was not a professional, was trained to pull teeth for people with toothaches. We were impressed with their dedication to helping people who otherwise had no access to medical care.

When we made our third visit to the Banks in 2005 we attended the festival at Vureas Bay, on the southwest side of Vanua Lava. Since there is very little in the way of accommodation in the Banks Islands, cruising sailors anchor their boats in the bays in front of the village, and row in to attend the festivities during the day, returning to their boats at night. Vureas Bay is adjacent to Veteboso, one of the largest villages in the Banks, with about 700 residents. It contains a small store, a church, and an elementary school. At the time we visited there were about 200 students attending the school, which covered grades one to eight. The only high school was at Sola, on the other side of the island, a half-day walk from Veteboso village.

The government does not provide free education in Vanuatu. Families must pay school fees for their children, costing about $10 per quarter, or $30 a year. There's very little way to earn money in the Banks Islands. Sometimes a cruiser is adopted by a family, and agrees to pay the school fees for the children until they graduate. Friends of ours gave money for a scholarship fund for needy students at the elementary school at Veteboso. They also helped one family by setting up the father with a small sawmill business so that he could earn money for his son's education. They later paid for the son to enroll in computer classes, leading to a successful career designing websites for the government.

On the last day of the festival at Vureas Bay in 2005 we were taken on a tour of the village, the church and the school. The school was a small two-room cinderblock building with a rusty roof. The teacher who was our tour guide told us that when it rains the roof leaks so badly that school has to close for the day. After the festival was over, Dan and I stayed behind to have a talk with the head teacher of the school. We asked what it would cost to buy materials to repair the roof, and agreed to deposit money in the school's bank account to pay for new corrugated iron panels for the roof. We wanted to do something to help all the children who attended the school rather than pay the school fees for a few children.

After we left Vanuatu on the way back to Australia we heard on the ham radio net that two young girls from Veteboso village, Marcella and Nelian, wanted to return to high school to finish their education. They had both been out of school for a couple of years. The boys in their families were being educated, but there was not enough money to pay for the girls' education. We knew both girls and their families, and had watched the girls mature from 10-year-olds to 15-year-olds on our visits between 2000 and 2005. They had served as guides on the trip we took to the proposed water generator site on the last day of the festival at Vureas Bay. Judy asked if anyone could sponsor them. We agreed to pay their school fees until they graduated. Bob and Judy on *Siddiqi* were about to leave Vureas Bay, and would take the two girls to Luganville, where they would stay with Jill at the Adventist School and learn to sew while they applied to schools of their choice. Marcella chose a Seventh Day Adventist school at Luganville and Nelian chose Ramwadi High School on Malekula. They both began classes within a couple of months. An Australian couple we knew who were also at the festival later agreed to share the expenses with us.

Although we didn't return to Vanuatu after our 2005 visit, the following year we decided to donate money into a school maintenance fund to be used however it was needed. We wrote to the new principle with our proposal, and he replied, giving us his school account number so we could transfer the money directly. A year later the principal wrote to us that they were using the money for a small library adjacent to the school. That year an American Peace

Corps worker had been assigned to Veteboso village, and he had helped in the construction of the building. We were delighted that they had used our donation for this purpose—the first library in the Banks Islands.

The next year we received a plaintive request via email from Cainton, the young student our friends had supported through computer classes in Luganville. His cousin, Mary Vira, was studying accounting at the Vanuatu Institute of Technology. After two years her family could no longer pay for her school fees. We agreed to pay her fees until she graduated.

Nelian and Marcella dropped out of school the second year and both became pregnant. Although they didn't finish school we hoped that the additional education made a difference in their lives. Mary Vira completed a further two years and graduated third in her class. She obtained a vocational certificate in accounting. We learned she had returned to her village to do the bookkeeping for her father who ran a small store in the village of Asanvari. We were proud of her accomplishments. In return for our wonderful visits with the friendly people in the Banks Islands we were happy to support the elementary school and the education of girls whose parents couldn't afford the school fees.

## Tropical Illnesses

Before we went cruising offshore we assembled an extensive medical kit. During passages we would have to treat any illnesses or accidents on board ourselves. Many of the small islands in the Pacific have minimal medical facilities, most often poorly supplied clinics. We made a list of medicines recommended for cruising sailors and took it to our doctor in Seattle. He wrote prescriptions for several antibiotics, such as Keflex, Amoxicillin, and Cipro, plus Silvadene burn cream, codeine, and cough syrup. In addition we bought over-the-counter supplies such as aspirin, an emergency dental kit, bandages of all sizes, forceps, gauze pads, tape, and elastic bandages for sprains. We bought several books on advanced first aid and collected articles on tropical diseases. We felt we were prepared for just about anything.

The cruising life is essentially a healthy existence. At sea we are away from the pollution of big cities, and breathe fresh, clean ocean air. On the boat we eat mostly vegetarian food, fresh fruit and vegetables and fresh fish when we can catch it. We are not tempted to eat huge steaks, pizza, hamburgers, or ice cream. It is difficult to get exercise by walking on the boat. Our decks are only 28 feet in length and floor space inside is only three feet wide by nine feet long. Sailing the boat gives us some arm exercise, cranking in the foresails with the winches, and pulling in the mainsheet when jibing. Dan does all of the sail changes and gets a great deal of exercise raising and lowering the sails and reefing and unreefing the mainsail. In the rolling seas we use our muscles constantly to keep our balance, particularly our abdominal muscles. We get most of our exercise while standing watches. During our three-hour watches throughout the day and night we climb up five steps to get from the cabin to the cockpit every 10 minutes to look around for ships and check our sails and course. This amounts to 48 times a day for each of us, equivalent to climbing a flight of stairs in a house about 20 times a day.

Once we reach land we walk everywhere—to buy and transport groceries, visit villages, hike to waterfalls, or to birdwatch. When at anchor we row the dinghy to get ashore, and swim and snorkel in tropical waters. In towns or cities we walk to restaurants, concerts, movies, and Internet cafes. It is a healthier and more active life than when we were both working in Seattle.

Many cruisers we knew had serious medical conditions before they went cruising including back problems, hip and knee problems, a triple-bypass, asthma, and atrial fibrillation. We were fortunate that we had no serious medical problems before we left. To make sure we stayed healthy every two years we flew back to the U.S. and had checkups in Seattle with our doctor, dentist, dermatologist and optometrist. They were preferred providers, covered by our health insurance.

Diseases encountered in the tropics include dengue fever, malaria and ciguatera. Dengue fever causes severe headaches, pain in the joints, a high fever and general crankiness. It is spread by a daytime mosquito. When we sailed to the Marquesas in 1989 many of our friends came down with dengue fever after visiting the island of Nuku Hiva. There was an epidemic on the island. When it hit cruisers after they left the Marquesas to sail to the Tuamotus it was particularly disruptive. The victim was so sick he or she could barely stand their watches. Their spouse or partner essentially became a single-hander, doing most of the work on the passage. To avoid dengue fever we decided to skip sailing to Nuku Hiva, sailing on to the next island, Ua Pou. When we sailed back to the Marquesas six years later there was no epidemic, so we were able to visit Nuku Hiva.

One cruiser we knew contracted malaria twice while visiting Vanuatu. She and her partner spent a year helping the people on the island of Erromango with various projects. They slept year round in a thatched hut with inadequate mosquito protection. Like most cruisers, we visited Vanuatu only in the winter, when there's little malaria. Nevertheless we always took a daily dose of doxycycline as a preventative.

Ciguatera was another hazard in the tropics. Several people we knew came down with ciguatera from eating large predatory fish in coral areas. Ciguatera poisoning causes vomiting and diarrhea, extreme lethargy, and a reversal of the sensations of hot and cold. The symptoms can last for months. Fortunately we avoided it. The fish we caught in the open ocean, like tuna, wahoo, and mahi mahi do not accumulate the ciguatera toxin.

The main problem we had in French Polynesia and Vanuatu was staphylococcal infections. A small cut, scratch, or mosquito bite could quickly develop into a serious staph infection. On our first trip to French Polynesia I developed a blister from my sandal strap rubbing on the back of my heel. Overnight it became red, swollen and painful. From reading our first aid books we learned that untreated staph infection could be life threatening, requiring hospitalization and treatment with drip antibiotic. I soaked my heel and elevated my foot until it cleared up. Years later when we visited Vanuatu both Dan and I occasionally developed staph infections on our legs or feet. If the infection didn't clear up in a few days we took a 10-day course of Amoxicillin, which always worked. Many of our cruising friends developed staph infections in Vanuatu and in Australia. We knew two cruisers, both men, who ended up in the hospital for treatment in Australia because they thought their leg infections would heal on their own.

During our years of cruising we heard about many serious medical problems of our fellow cruisers. One cruiser developed pain in the right side of his abdomen on a passage from Australia to Vanuatu. He spent most of the passage lying in his bunk while his wife and a crewmember sailed the boat to Vanuatu. It turned out to be diverticulitis, and he soon recovered after reaching port. We heard the report from a single-handed sailor on the way to Fiji who had severe pain from a kidney stone. He was told to drink plenty of liquids on the boat. When he reached Fiji he drank lots of beer and finally passed the stone, to his great relief. A cruising friend had a sharp pain in her groin, nausea, and vomiting while they were anchored at an island off the Queensland coast of Australia. Her husband managed to get her ashore to the island in their dinghy, where she was airlifted by helicopter to the mainland hospital. She had surgery the same day for a femoral hernia.

We had only two medical problems that required seeing a doctor in a foreign country. The first was Dan's kidney stone, which occurred when we were in a marina in New Zealand in 1997. The second was a small lump on my upper back. I noticed it a few weeks before we left on our first trip from Australia to Vanuatu in 2000. During the passage it began to enlarge from the size of a pea to about the size of a walnut. Two months later on the passage from Vanuatu to New Caledonia it became painful because it was irritated by the straps on my safety harness. In Noumea we visited a French doctor who was recommended by the staff at the marina office. The doctor spoke no English, but with my limited high school French I could understand a few words. He explained that what I had was a blocked sebaceous gland. He could treat it, but couldn't guarantee that it wouldn't come back in three or four years. Dan watched while the doctor cut a half-inch slit in the top of the lump, and squeezed out the sebum. The doctor stuffed the hole with a long roll of gauze, leaving an inch sticking out. Then he covered the area with a large bandage. Since I couldn't reach the area, Dan had the job of applying antibiotic cream and changing the bandage daily. After 10 days Dan pulled out the gauze, and the problem was cured.

Considering all the diseases and medical conditions we could have had, we felt extremely fortunate that we came through relatively unscathed. Our medical problems occurred on land where we could get medical care. Being cautious and lucky, we escaped most of the tropical diseases that plague cruisers. Because of the active lifestyle we probably stayed healthier during our years of cruising than if we had stayed home in Seattle.

## A Dangerous Infection

In 2005 we visited several of the small remote islands of the Banks Islands in northern Vanuatu for a month, and now we were anchored at Vureas Bay, waiting for Gordon and Miriam to join us. We planned a last visit with Godfrey and his family before leaving the Banks Islands and heading south to Luganville. The cruising season was nearly over, and we were planning to head back to Australia from Luganville. Gordon and Miriam were anchored at Waterfall Bay, where Gordon was suffering from a swollen and painful knee. Just after he anchored he had kneeled with one knee on the boom gallows in order to put the sail cover on his mainsail, and experienced searing pain and swelling just below his knee. We and the crew of seven or eight other boats who were scattered around Vanuatu and Australia had a twice-daily ham radio schedule, and Gordon reported for several days that he was feeling ill, and had a fever. It was fairly low in the morning, but spiked to 103 or more in the afternoon. Since a fluctuating temperature is a symptom of malaria, we and other cruisers advised him to take malaria pills, which most of us had in our medical kits as a precaution. There are no doctors in the Banks Islands, so everyone has to rely on their medical kits for any problems that arise. Unfortunately Gordon felt worse after taking the recommended two-day dose of malaria pills. We now suspected he had a serious staph infection in his knee, even though there was no broken skin. He had had an infected mosquito bite on his ankle 10 days before, but he thought it had healed after taking a week's dose of antibiotics. Three days after taking the malaria pills Miriam reported on the net that Gordon had had a bad night, and his fever had spiked to 104.5 in the afternoon. We were very worried about him. We decided to sail back north to Waterfall Bay to try to help. Our friends, Mike and Sue, on *Yaraandoo II*, who were anchored near us, made the same decision. As soon as it was light the next morning we both raised anchor and made a quick sail to Waterfall Bay, arriving about 9 am.

After we anchored, the four of us rowed over to Gordon and Miriam's boat, and sat around in their cockpit discussing how to best get Gordon to Luganville, where there are doctors and a hospital. That afternoon we were all surprised when some of the villagers from Vureas Bay arrived on the shore at Waterfall Bay. They had hiked for several hours on the rocky path between these villages. The party consisted of Godfrey with his wife Veronica, and Swithen, his wife Helen, and their daughter Marcella. Carolee, who lived at Waterfall Bay, ferried the group of villagers to Gordon and Miriam's boat in his outrigger canoe. Veronica tried folk medicine, gently massaging Gordon's red and swollen knee for an hour. Unfortunately this did nothing to reduce the swelling or relieve the pain. They promised to return the next day to apply a leaf, which we assumed was a kind of folk remedy. We all thought it was best to get Gordon to see a doctor as soon as possible.

Early the next morning Dan helped Miriam raise the sails and pull up the anchor on *Anwagomi*, then we raised our anchor and sailed along with them for the 90-mile trip to Port Olry on Espiritu Santo Island. Our friends Mike and Sue also left at the same time, but planned to take the five villagers back to Vureas Bay, so dropped way behind us. Miriam sailed their boat essentially single-handed, with Gordon lying on the berth below unable to go out on deck to help. The wind was about 30 knots against us, so it was a rough trip, sailing against the wind, assisted by the motor when necessary. Miriam did a great job of handling their boat. By the time we arrived at the pass between a small island and the large island of Espiritu Santo it was midnight. This was about nine miles north of where we planned to anchor at Port Olry. As a short cut we planned to go through the mile-wide but uncharted pass into the anchorage, but Dan and I had never taken this route. We preferred to go around the outside of the small island to avoid the uncharted pass. Gordon and Miriam had navigated this several times before, but always in daylight. Because they had experience we followed them through the pass, Miriam using radar to find the way in pitch darkness. As we approached what we thought was the pass Miriam suddenly reversed course. She called us in a panic on the VHF radio, saying, "We nearly hit a reef. I saw breakers a boat length ahead of us. We just made a 180-degree turn and are headed back north." We quickly turned our boat around and headed north into deeper water. This was very frightening. If we had continued on, both boats would have gone aground and possibly wrecked on the rocks. It turned out that the chart was off by about half a mile, and their radar was malfunctioning, with the right half of the screen being blank. Miriam mistook the blank part of the screen as the pass. It took us a while to calm down and assess the situation. We were in a real quandary. Our radar was working, but we didn't think we had properly aligned it when it was new, so we didn't trust it. Fortunately our friends Duncan and Audry, who were anchored at Port Olry, heard our conversation on the VHF radio. Duncan could detect our two boats with his radar and locate us on his electronic chart. He gave us waypoints to lead us into the anchorage. He assured Dan and me that if we stayed in the clear area between the two landmasses on our radar screen we could make it through the pass. So we proceeded, with Miriam and Gordon following close behind, using our radar and Duncan's waypoints. We finally made it into the bay and anchored at 2:30 am. Everyone was totally exhausted and went to bed immediately.

At 6:30 am Miriam called Dan on the VHF radio and asked if he could take her ashore so she could find transportation to the hospital in Luganville. Miriam right away found a van parked near the anchorage that was leaving for Luganville shortly. Dan called friends in the anchorage using his hand-held radio to ask if they could use their large dinghy to pick up Gordon from his boat and take him ashore to the waiting van. Two men helped Gordon into a large inflatable dinghy. Dan accompanied Gordon and Miriam in the van for the two-hour trip to Luganville.

They reached the hospital at 9:30 that morning, and Gordon was seen by two doctors, admitted to the hospital and placed on drip antibiotics and painkillers for his infected knee. Dan and Miriam returned to the anchorage at Port Olry by taxi that afternoon. The next day Miriam went back to Luganville by truck, and moved into the extra bed in Gordon's room to help with his care. For the next week Dan and I alternated with Mike and Sue, taking 2-hour truck rides to visit Gordon and Miriam at the hospital. Mike and Sue had us over to dinner on their boat when we returned around 5 pm, and we did the same for them. After two days, with little improvement in his condition, the Fijian doctor told Gordon that he would have to perform surgery to remove the dead infected tissue, and irrigate the wound with antibiotic. Both Gordon and Miriam were understandably apprehensive about having surgery in this rather primitive hospital, so we called friends, Peter and Ruth, who were anchored across the channel at Aore Island. Dan asked them if they could visit Gordon and give a second opinion. Peter was a retired ear, nose, and throat doctor from Sydney. He took the next ferry across to Luganville to see Gordon. Peter assured him that this was the best procedure. That afternoon the doctor performed surgery, cultured the fluid from Gordon's knee, and found that the bacterial infection was pseudomonas. They switched to the appropriate antibiotic to treat the infection, and Gordon began to improve immediately. Two days after the surgery we were greatly relieved to see that he was back to his usual cheerful self, and able to walk around without pain.

After a week Gordon was able to leave the hospital. He and Miriam moved into a small bungalow at a resort near the waterfront. Gordon had to return to the hospital every morning to have the wound re-bandaged.

A few days later Miriam returned to *Anwagomi* in Port Olry. An old cruising friend, Kurt, helped Miriam sail *Anwagomi* 25 miles to the next anchorage at Peterson Bay. We sailed *Shaula*, and Mike and Sue sailed their boat, *Yaraandoo II*. The three boats arrived at Peterson Bay in late afternoon. This was much closer to Luganville. Miriam returned to Luganville by truck, while we stayed at Peterson Bay for several days. From Peterson Bay we had a one-hour truck ride into Luganville to visit Gordon and Miriam. The second leg, from Peterson Bay to Luganville was only 17 miles. Dan helped Miriam sail *Anwagomi* to Aore Island across from Luganville, then Dan returned to Peterson Bay by van. The next day Dan and I sailed *Shaula* down to Luganville and anchored in the bay in front of the resort. For the next week we remained anchored in this bay, and visited Gordon and Miriam every day.

During Gordon's hospital stay and recovery he and Miriam had lots of visitors and support of cruisers and villagers from Vureas Bay. To show their appreciation Gordon and Miriam invited a dozen of us to lunch at the small cafe in town that had delicious hamburgers and milk shakes. We all celebrated Gordon's recovery from a serious infection which, if left untreated, could have been fatal.

Most cruisers are prepared for emergencies at sea with books on first aid and tropical diseases, and an extensive medical kit with all kinds of antibiotics, painkillers, burn cream, and bandages, hoping that they will never have to use them. Many cruisers check into a daily radio net where they can get good medical advice while at sea or in port. It is comforting to know that if anything happens while in port, no matter how remote, there will be fellow cruisers willing to do whatever it takes to help out. We were glad that in spite of our misdiagnosis, and inexperience in navigating into a tricky anchorage at night, we were able to help Gordon get the medical care he needed. Many people in this close-knit cruising community contributed to this story having a happy ending.

Potluck on the beach in Sea of Cortez, Mexico

*Shaula* at anchor in Sea of Cortez, Mexico

Dan with rockfish and two lobsters in Sea of Cortez, Mexico

Pineapple fields on Moorea, French Polynesia

Cathy on our hike from Hana Vave Bay to Omoa, Fatu Hiva

Family on the trail to Omoa carrying copra on horseback

Main street at Hana Vave, Fatu Hiva

Dan and Bobby climbing on the trail over the top of Rarotonga

Hikers Jeff, Greg, Alice, Dan, Hanna, and Joann riding back to harbor in Telecom truck on Rarotonga

Jeff and Cathy climbing coral cliff at Togo Chasm in Niue

Jeff and Cathy in their homebuilt two-part dinghy

Dan and Jeff at Matapa Chasm in Niue

Dan on *Shaula* entering Opua, New Zealand with broken boom

*Shaula* anchored at Opua with broken boom on deck

Cris and Dick preparing sushi at their home on Kawau Island, New Zealand

Brian and Sue with their cat on *Waitoa*, Bay of Islands, New Zealand

Guide at Cascades on Efate Island, Vanuatu with Stan, Dan, and Mari

Dan entering the New Caledonia cultural center in Noumea

Dancers with fish headdresses at festival at Sola, Vanuatu

Men pounding taro for laplap at festival at Sola, Vanuatu

Interior of *Shaula* at anchor at Vanuatu. Vanuatu flag on starboard side

Dan, Jan, Austin, Lynn, and Jim enjoying Thanksgiving dinner at Bush Turkey Restaurant, Coffs Harbor, Australia

View of city skyline from our anchorage at Blackwattle Bay, Sydney Harbor

Kangaroo feeding time at Currimbin Wildlife Refuge, Queensland, Australia

Koala at Fleay's Fauna, Queensland, Australia

Site of the waterfall generator at Waterfall Bay, Vanua Lava, Vanuatu

Villagers at Waterfall Bay celebrating the installation of the waterfall generator on Vanua Lava, Vanuatu

Dancers at Bannan Bay, Malekula Island, Vanuatu

Dancers with drum accompaniment at Vureus Bay Festival, Vanua Lava, Vanuatu

Miriam, Gordon, Mike, Sue, and Dan
on the beach at the Reef Islands

Women on Guau Island making water music

Chief Godfrey with Sue and Mike at Vureus Bay

Cruisers relaxing in front of Chief Godfrey's guesthouse during the Vureus Bay festival

Richard on guitar and Alice on fiddle joining the string band at Luganville, Espiritu Santo

The wreck of British sailboat *Coker Lady* on the shore of Tanna Island outside of Resolution Bay

Feast prepared for cruisers at the Tanna Yacht Club at Resolution Bay

String band performing before the feast at Tanna Yacht Club

Snake dance during the dance festival on Tanna

Farmer's market at Lenakel, Tanna Island

Dan with string that tangled around our propeller shaft on our entrance into Noumea, New Caledonia

Alice's violin in three pieces after overheating in *Shaula*'s quarterberth at Lake Macquarie, New South Wales

Dockwise ship docked at terminal up the Brisbane River

*Shaula* on board the Dockwise ship in Brisbane

Alice on a hike in Joshua Tree National Park while *Shaula* is being shipped

*Shaula* on deck of Dockwise ship on arrival in Ensenada, Mexico

Interior of *Shaula* with everything from the deck stowed below, ready for trucking to Olympia

Loading *Shaula* onto truck for transporting to Olympia

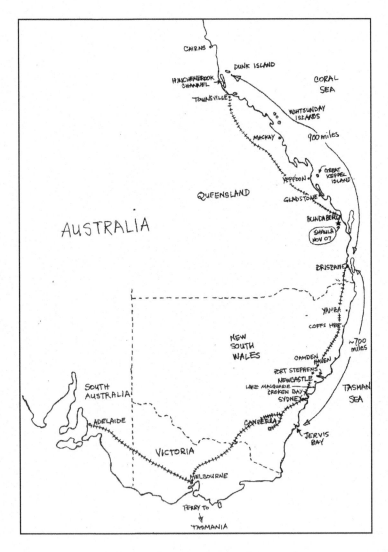

Exploring the East Coast of Australia

# 11 EXPLORING AUSTRALIA

## Immigration Troubles

After I was told to leave Australia for overstaying my six-month stay, Dan and I flew to New Zealand so I could apply for a new visa. It was traumatic to leave Australia not knowing if I would be permitted to return. After obtaining my new visa and returning to the boat we finally decided to apply for four-year retirement visas, which didn't require us to leave Australia after every six-month period. Australia was the only country we visited which required us to obtain a visa outside of the country. Retirement visas were an exception to the rule. It was possible to apply for and obtain this visa while living in Australia. We were planning to fly home in mid-June 2004 for three months, so we wanted to apply for the retirement visa before we left.

The process started by our picking up the application forms at the immigration office in Brisbane in mid-May 2004. For the next month we went through an amazing number of hoops. First we made appointments for medical examinations. The examinations consisted of a blood draw to check for HIV, and a chest X-ray the next day. After we passed those tests we had to go for a basic physical exam. The doctor checked our eyesight, blood pressure, heart, lungs, and our physical condition, by having us touch our toes, and squat down. While we were in downtown Brisbane we went to the police department to have fingerprints taken, to be sent to the FBI. We had to pay $170 each for the physical exam, and $30 for the fingerprints. The 20-page application form took a day and a half to fill out. It required listing every country we had visited in the last 10 years, proof we had medical insurance, a character assessment questionnaire, and copies of our investment and bank statements to show we were financially self-supporting. Then we had to have passport photos made to attach to the application. Finally, a week before we flew from Brisbane to Los Angeles we mailed the bulky applications to the immigration department in Perth. The day before we left we called the immigration department to make sure they had received them. We told them we would be out of the country, but they could contact us by email.

Obviously, we had to really want to stay in Australia in order to go through all of this. Australia was a wonderful place to cruise, with mild weather that allowed us to sail year round. The people were friendly everywhere we went, and we enjoyed the company of many Australian and international cruising sailors. By the time we applied for retirement visas we had been in Australia for four and a half years. We had settled into a pattern of sailing to Vanuatu and New Caledonia one year, then flying home the next year. Our visitor's visas only allowed us to remain in Australia six months without having our passports stamped in another country. During our six months in Australia we cruised the coast between Bundaberg and Sydney. It left us with little time to cruise more of the coast, or travel to other parts of Australia. We were looking forward to having the freedom to stay longer so we could explore more of the country.

After we returned from our trip to the States in 2004 we began our migration down the coast from Brisbane to Sydney. We planned to go to the Sydney office to have the visas put into our passports. We could be in touch with the immigration office in Perth by phone or email in all the small towns along the way as we cruised down the coast.

Our troubles began in February, five months after we returned from the United States. We had heard nothing from immigration for eight months, except that they had received our application in June. The immigration office had assigned us a caseworker who was supposed to process our application, but we were told to not call the caseworker unless it was absolutely necessary. The caseworkers were extremely busy, and calls would delay the process. But the six months we were allowed to stay in Australia was running out. If we didn't receive the approval by mid-March we would have to leave the country or be deported. We couldn't sail to Vanuatu in March because it is still cyclone season. Having just returned from the States in September, we weren't eager to fly home either. Because of the high cost of plane tickets we usually flew home only every two years. Not knowing what was causing the delay, I finally went to a pay phone to call our caseworker in Perth. She would only say that she was working on it, but she had a large stack of applications.

Then, after talking to friends who had received retirement visas a year earlier, we figured out the problem. Our friends had signed up for Australian health insurance, because their U.S. health insurance wasn't adequate—it had to cover 90% of medical costs in Australia. We checked our U.S. health insurance policy, and found that it only covered 85% of medical costs when out of the country. We called Perth again to confirm that we needed to buy Australian health insurance, and found out our caseworker was no longer there; we had been assigned a new caseworker. The new caseworker told us that we must have Australian health insurance. A mad scramble began. Researching different health insurance companies on the Internet we narrowed it down to two companies. I called immigration again to make sure these would meet their requirements. By this time a third caseworker, Daphne, had been was assigned to our case. Daphne informed us that the Australian health insurance must have the word "healthy" in its title. She also said she wouldn't look at our file until all the paperwork was in. Then it would take a month to process, because there was a long queue ahead of us. It was now February 18th, so we didn't have a month before we had to leave the country.

We were getting desperate.

We called Australian Unity in Melbourne, signed up and paid for a year of Australian Unity Healthy Travelers Insurance. This wasn't easy. They required a copy of our passports showing the date of last entry into Australia, and an address of an Australian resident to whom they could send our insurance cards. After we gave them this information it took a week for the insurance company to send us confirmation for the immigration department. Time was running out. We explored the possibility of obtaining a bridging visa, but that is only good for two months. While on a bridging visa a person is considered to be a temporary resident, and customs requires posting a bond on the boat (for one year) or importing the boat. This would cost a great deal of money. We'd already paid $500 for the applications and medical fees, and $760 for Australian insurance for one year. We still didn't know for sure if the immigration department would grant us retirement visas.

A few days later I called immigration in Perth once more. Our caseworker was out for her lunch, but the woman who answered the phone, Rudecka, listened to our plight. She was sympathetic, and thought it would save a great deal of work if they could finalize our retirement visas before our visitor's visas expired. She said she would take over our case if the only outstanding item was confirmation of health insurance. She said she'd try to finalize the retirement visa by March 10$^{th}$. Hooray! Finally, we had a caseworker who cared. We were delighted when we went to the post office March 3$^{rd}$ and found our approval letter from immigration in the mail. All that was left to do was sail the 15 miles from Broken Bay to Sydney, hurry down to the immigration office, and have our retirement visas pasted into our passports!

There were many times over the nine months of waiting when we wondered if it was worth all the hassle and money to obtain these visas. But now we didn't have to worry about immigration breathing down our necks for the next four years. We could go anywhere we wanted in Australia at our own pace. It gave us a wonderful feeling of freedom.

## Exploring by Land

Tasmania, Australia's smallest state, was the first place we explored by land. Many of our cruising friends visited Tasmania by boat, some sailing there every year. From Eden in New South Wales to Wineglass Bay in Tasmania is more than 500 miles across Bass Strait. This body of water can be extremely rough, because the winds and seas are unimpeded as they circle the globe in the roaring forties. Often sailors wait for days in Eden for a favorable weather window. In our boat the trip across Bass Strait would take four to five days. Even if the weather is good at the beginning, adverse winds or storms can come up suddenly, kicking up dangerous seas. We wanted to avoid this potentially rough voyage.

After returning from our plane trip to New Zealand in 2004, we decided to visit Tasmania the easy way. Our boat was safely tucked away in a marina in Brisbane. From there we took an eight-hour train trip to Sydney, and the next day a 10-hour train trip to Melbourne. The trains in Australia are fast, comfortable, and pleasant. Crossing Bass Strait by ferry was an easy 10-hour trip. Once in Tasmania we travelled to many coastal and inland cities by bus, called Tassie Link. The bus rides were usually only two or three hours, as distances between towns are short. We were able to book accommodation at inexpensive backpackers and hostels a day ahead of our arrival.

About a quarter of the land of Tasmania is devoted to national parks. We chose to visit five of the 19 national parks: Freycinet, Maria

Island, Mt. Field, Cradle Mountain, and Lake St. Clair. We took a day cruise on Lake St. Clair, the deepest lake in the southern hemisphere. We enjoyed many day hikes on the wonderful hiking trails in all of the parks. We were able to see some of Tasmania's impressively tall eucalyptus trees, such as Tasmanian Blue Gum, and Mountain Ash, which can grow to over 300 feet tall.

In Hobart, the largest city and capital of Tasmania, we took a tour of the Cadbury chocolate factory, where we were given samples of many varieties of their delicious chocolate candy. We visited the Wooden Boat Centre school in nearby Franklin, where they were building wooden versions of our boat, and the Apple Heritage Museum, with its hundreds of varieties of apples. In Launceston we attended a wonderful free concert in Royal Park by the Tasmanian Symphony Orchestra. At Triabunna, we visited a huge wood-chipping facility. At the backpackers where we stayed huge trucks filled with large diameter logs passed by all day long. The owner of the backpackers told us that the Gunns Limited forestry company was chipping 7,000 tons of old growth logs per day, and shipping three boatloads of woodchips a week to Japan. There were many protests in the forests against logging of old-growth timber. A few hardy souls stayed in the tops of some of the tallest trees to prevent them being cut down. Most protesters were simply ignored by the forestry industry.

From Strahan, on the west coast of Tasmania, we took an all-day excursion on a catamaran from Strahan Harbor to the Gordon River. We stopped at Sarah Island, where convicts harvested huge stands of Huon pine in the 1820s. They built 130 ships over a period of 25 years. The last ship they built was stolen by 10 convicts and sailed to Chile, where it was scuttled. Years later the convicts, who had settled in Chile and married Chilean women, were charged and tried in Hobart for stealing the boat. They were let off, however, because the ship was never registered, and therefore never existed. The last stop was at a Huon pine log mill, where valuable Huon pine trees that had survived decades underwater were being milled for furniture making. The water-repellant oily yellow wood is a superior boat-building material. The few Huon pines still growing in Tasmania are preserved from logging.

On our visit to Freycinet National Park, we hiked a trail over large boulders to Wineglass Bay. At anchor were our friends Frank and Teresa, on *Dansa na Marra*, from Scotland. It was their first anchorage after crossing Bass Strait. We enjoyed learning about their adventures of crossing the strait. Other cruising friends, Jim and Ann, met us at the Salamanca market in Hobart. Their boat, *Insatiable II*, was anchored in the Derwent River. They invited us aboard for lunch and we enjoyed an afternoon sail on their new 40-foot Australian-built wooden boat, and dinner at an Indonesian restaurant in Hobart. An old cruising friend, Carl, from our days of living in Sydney in the 70s was working in Hobart. We were able to visit him and his partner Chris in their beautiful home south of Hobart.

We returned by ferry and train to Brisbane after our six-week tour of Tasmania. Tasmania was a fantastic place to visit. We were glad we were able to travel there without sailing across Bass Strait in our boat.

After our long-awaited retirement visa was finally granted in March 2005 we decided to see more of Australia at our leisure. In February 2006 we took a day train from Sydney to Melbourne, and stayed there a week. Melbourne is one of the largest cities in Australia, with a population of 4.2 million. We stayed a few days at the home of cruising friends, Grant and Lee Ann, who took us on great hikes in the surrounding hills and mountains. They also took us on a wonderful drive along the Great Ocean Road, where we took walks down to the ocean to see the spectacular rock formations just offshore. The city of Melbourne has an extensive tram system which allows you to travel just about anywhere in the city. The tracks are in the center of the street, with cars traveling on either side. The trams come by every 10 minutes, so there is only a short wait. We took a tram to the harbor where the Round the World racing boats were tied up. The skippers and crew were working frantically to get ready for the next leg of the race. They were focused on getting to the next port as fast as possible, pushing the boat and themselves to the limits of their capability. Visitors were welcome at the dock, and we learned about the equipment, food, and courage needed to sail in this treacherous ocean race. It was in sharp contrast to our cruising, where we avoided storms and suffered few hardships as we cruised at our leisure in the relatively benign Pacific Ocean.

Our next stop, Adelaide, was a long overnight train trip from Melbourne. For us, the main attraction in Adelaide was the wineries. We spent an afternoon at the fine University of Adelaide wine museum, where we learned how they grow grapes, and make red, white, and rose wine. They had a wine-tasting bar where we could taste five different Australian wines. Later we booked a wine-tasting tour, which took us to the hilly country around Adelaide where some of the best wines were made. The luxurious bus had only 12 other passengers. The tour stopped at four of the premium wineries around Adelaide. Not only did we get to taste some amazing wines, we were given tours of the vineyards, and learned about the growing of grapes, and making and aging of wines. We resisted buying any cases of wine. Where would we put a case of wine on our small boat? After sampling many different varieties, we decided that Shiraz was our favorite.

Our last inland trip in Australia was a visit to Canberra. Australian friends offered us the use of a mooring buoy in Broken Bay, where we could leave our boat for free while we traveled inland. It was an easy four-hour train trip from Sydney to Canberra. We stayed at the youth hostel in central Canberra for a week. Parliament wasn't in session, but we enjoyed taking a tour of the parliament building and learning about their system of government. The capitol is an attractive town with a small-town feel. The streets are planted with two million trees from all over the world. Our visit was in April, the fall season in Australia. We saw an amazing display of fall color in the trees. This is a rare treat in Australia where most of the trees are eucalyptus, which stay green year around. We enjoyed crunching on acorns on the sidewalks of Canberra as we strolled around town. We visited the War Memorial Museum and the botanical gardens, and watched a display of hot-air balloons drifting over the lake in the center of town.

Our inland trips seemed much less challenging than our travels on our boat. We could cover long distances in comfortable air-conditioned trains, buses, and trams. We saw scenery of the Australian outback, and could relax and talk with fellow passengers. For a change we were traveling like "normal" tourists, and found it had many advantages over sailing in our small boat in rough seas.

## A Musical Interlude

Our trip home in 2006 to visit Dan's mother in California and friends and relatives in Seattle was a short one—only two months, from mid-March to mid-May instead of our usual five or six months. Dan's mother had moved to a retirement home in 2002. She no longer had to keep up a house and garden or prepare meals, so didn't need our help with household chores.

We had purchased tickets for the Australian Festival of Chamber Music in Townsville, Queensland, which started in July. Our plan was to sail up the Queensland coast to Townsville, and stay at the marina in order to attend the festival. We were looking forward to a week of chamber music while staying on our boat.

One of the things we missed while cruising in Australia was attending concerts of classical music. In Seattle it was easy to attend concerts. We had enjoyed going to Seattle Symphony concerts, chamber music concerts at Meany Hall, and the annual Seattle chamber music festival at Lakeside School. While cruising in Australia we spent most of our time in small towns along the coast. The only opportunity we had to go to concerts was when we were in Sydney or Brisbane. The Townsville chamber music festival would make up for sparse concert experiences we had in our eight years of cruising in Australia.

Unfortunately the weather in June in Brisbane was miserable. Day after day the rain pelted down, along with strong winds. After three weeks waiting in the marina in Brisbane for decent weather, we realized we couldn't sail to Townsville, over 800 miles to the north, in time for the festival. Not wanting to give up on attending the festival, we decided to sail about 150 miles to Bundaberg, and take the train or bus to Townsville. We made reservations to stay for a week at the Civic Guest House in Townsville. The trains only traveled three times a week, so to get to the festival on time we had to take a bus. We made reservations for an overnight bus leaving the evening before the festival.

The bus left at 11 at night for the 15-hour trip to Townsville. We appeared to be the only passengers over 30. The bus was filled with a young crowd of backpackers headed for the Queensland beaches for the school holidays. As the bus left the station two drunken English girls staggered up and down the aisles, laughing wildly and making loud remarks that could be heard throughout the bus. The bus stopped at several small towns along the way where we could get out and stretch our legs. At 5:30 in the morning the bus stopped at a small cafe for an early morning breakfast. There was only one toilet behind the cafe, which meant waiting in a long line. The stop for lunch was at another small cafe, where we had sandwiches. After a night of very little sleep it was a relief to at last arrive in Townsville at around 2:00 in the afternoon. The bus driver called the manager of the guesthouse, who picked us up at the station. Our room at the guesthouse was small, but comfortable, and we were close enough to the festival events that we could walk to most of them.

For the next six days we luxuriated in a feast of music. The first concert was in the evening of the day we arrived. It was given in Queens Park, where comfortable chairs were set up on the lawn. The concert was free, in order to attract the local Townsville residents. The audience of about 1,000 people was made up of mostly middle-aged adults or seniors. The director had a special treat—he had brought members of the Janacek Philharmonic, an orchestra that he conducted in the Czech Republic, to Townsville to perform as a chamber orchestra. They played an impressive program of Dvorak, Beethoven, and Mendelssohn's violin concerto. The soloist was a young Canadian violinist who played beautifully.

Every morning we attended lectures on such diverse topics as Shostakovich, composing in Australia, and the influence of Indonesian and Japanese music on Australian compositions (one of the speakers played the shakuhachi, a Japanese flute). In the afternoons we observed master classes or attended emerging artist concerts. The emerging artists were young talented Australian musicians just starting their musical careers. The evening concerts consisted of string quartets or quintets playing wonderful chamber music. They were preceded by Reef Lectures—so called because Townsville is in Northern Queensland, inland from the Great Barrier Reef. One lecture was on management of coral reefs in Indonesia, one on the spread of infectious disease by plane travel, and another consisted of images of diving humpback whales on a large screen with a background of whale songs. The Janacek Philharmonic musicians played a piece by Australian composer Peter Sculthorpe for cello, didgeridoo, and orchestra. One of my favorite quartets, Dvorak's American Quartet, was played by four young musicians. At the master class I had heard the group being coached by Lara St. John, a world famous Canadian violinist. It was amazing to hear how much they had improved after that coaching session. When one quartet performance was cancelled due to illness of one of the players, Lara St. John filled in by playing a Bach suite, accompanied by a didgeridoo for the last movement (a strange combination). The final concert was given in a cathedral on Sunday morning, and consisted of Mozart's Oboe Quartet and a Brahms piano quintet. The whole festival was a wonderful experience, both educational and entertaining. It far exceeded our expectations.

We had reservations for a train to take us back to Bundaberg. They served a delicious dinner and an early breakfast in the dining car. It was a long ride, 4 pm till 9 am (17 hours), but much more pleasant than the bus trip.

## Extended Coastal Cruising

Our retirement visa gave us the chance to sail along the Australian east coast at our leisure. Our usual coastal cruising was limited to the stretch of coast between Bundaberg in Queensland and Sydney in New South Wales. Now we had time to sail north to the tropical waters of Northern Queensland.

Cyclones are a danger in Queensland in the warm summer months from December to March, so we planned to sail as far north as we could during the winter and spring. In July, after we returned from the Townsville chamber music festival, we began preparing for the trip north. Fortunately our friends Gordon and Miriam from

*Anwagomi* sailed into Bundaberg at this time, and we decided to sail the Queensland coast together. Their boat, a Golden Hind 31, is comparable to our Bristol Channel Cutter 28 in size, making it possible to travel at approximately the same speed.

The Queensland coast north of Bundaberg offered some wonderful cruising in mostly sunny weather and light southeast winds. There is no ocean swell due to the protection of the Great Barrier Reef, which runs along the coast about 50 miles offshore. The currents off the Queensland coast are weak compared to the strong coastal currents that had plagued us as we cruised along the New South Wales coast. This is some of the best sailing in Australia.

The Whitsunday Islands were a big attraction. They are a group of islands just off shore with many protected anchorages, white sand beaches, wonderfully clear water, and resorts that welcomed visitors from yachts. People come from all over the world to charter sailboats for a week or more in the Whitsunday Islands. We enjoyed the beautiful tropical beaches, swimming in the warm waters, seeing humpback whales, and hiking on sandy trails to viewpoints overlooking the ocean.

There are protected anchorages that are less than 40 miles apart all along the coast and on adjacent small islands. We could sail from one to another during daylight hours. We enjoyed nearly deserted beaches at Scawfell Island and Curlew Island. Marine parks at Great Keppel Island, Magnetic Island and Dunk Island all offered well-maintained trails for hiking and observing the colorful tropical birds and interesting eucalyptus forests. A highlight of the trip was Hinchinbrook Channel, a narrow channel between Hinchinbrook Island and the mainland, with spectacular mountainous scenery. After a pleasant sail in the flat waters to a beautiful anchorage, we would get together with Gordon and Miriam to explore the beach or take bird walks on inland trails. Often we exchanged happy hours before dinner on either *Shaula* or *Anwagomi*.

At Yeppoon, Gladstone, Mackay, and Townsville we stayed at marinas for a week or more. When we had periods of rain and strong winds we were glad to shelter in the marina waiting for the weather to improve. Cruising friends who had settled in these small towns invited us to visit them ashore, and we also enjoyed occasional meals at Chinese or Thai restaurants.

Altogether we stopped at 27 anchorages, mostly on small offshore islands, and stayed at seven marinas as we sailed north. In October it was time to turn around to head south before the hurricane season hit. For a week we had strong southerly winds and rain while we were anchored in Hinchinbrook Channel. Finally the winds shifted to northerly and we were able to make good progress sailing south. We stopped at some of the same islands we visited on our northward trip, but also made an overnight sail, and a two-day sail while the winds were favorable.

It had taken three months to sail 900 miles to Dunk Island, but only two months to sail back to Brisbane. We were certainly glad we had obtained retirement visas so that we could stay in Australia for more than six months at a time. This allowed us to take this extensive cruise of the Queensland coast. With its small towns, many miles of beaches, and small offshore islands it is a beautiful part of Australia.

## Dunk Island

It was the last leg of our trip up the Queensland coast. We had been traveling up the coast from Bundaberg for the last three months with our friends Gordon and Miriam.

We woke to another clear, sunny day, looking forward to a day sail to

Dunk Island, about 25 miles north of where we were anchored. At our usual speed this would take five to six hours, so we needed to get an early start to arrive at our destination before sunset. After a quick breakfast we began checking for wind, even a light breeze so that we could start sailing. The wind was not cooperating, however, and at 10 am we decided we could wait no longer. We started the engine, raised our anchor and began motoring north.

After an hour of motoring we felt a light breeze coming up, so Dan raised the mainsail and we rolled out the jib, turned off the engine and began slowly sailing in flat, calm seas. Dan is impatient sailing in light winds, but I gladly took the helm, enjoying the challenge of trying to make the boat move as best I could under light air conditions. Dan went below to relax and read, and wait for his turn at the helm when we had more wind.

Unfortunately the slight breeze soon died, and we were just drifting with 20 more miles to go to Dunk Island. I decided it would be best to motor again for maybe 10 minutes to a patch of wind on the water just north of us, hoping we could then resume sailing. I started the engine and put it in gear, but within a few minutes we heard a terrible crunching, grating noise coming from beneath the boat. Dan immediately shouted up to me to put the gear in neutral, which I did, and thankfully the noise stopped. This was no ordinary engine noise. It sounded like the chopping up of a branch in a wood chipper. Had we run over a log or a tree branch that was being chewed up by the propeller? We didn't see any debris in the water or any floating branches.

Once again I rolled out the jib to resume sailing in the light breeze while Dan checked out the problem. With no engine power, and winds so light we were barely moving through the water, I worried that if a ship appeared on a collision course with our boat we would be like a sitting duck, unable to move. Dan checked the engine compartment. Seeing nothing amiss with the engine, he crawled through the small space beside the engine to get to the aft part, where the transmission shaft is connected to the propeller shaft. Ah, mystery solved! What he saw was a six-inch-long irregular bite taken out of the 3/8-inch plywood platform behind the engine. A pile of chips and sawdust sat next to the muffler, and three bolts were strewn in various places around the engine compartment. He quickly figured out what had happened. The four bolts that held a plate on the coupling between the transmission shaft and the propeller shaft had all fallen out, allowing the propeller shaft to shift aft and strike the plywood while spinning at about 700 revolutions per minute. Two longer bolts stuck out vertically on opposite sides of the shaft, and as it rapidly revolved they had chewed chunks out of the plywood, making a horrible din. Dan was able to retrieve the three of the bolts and bolt the whole thing back together. He finished the job with a fourth bolt, slightly shorter, that he found in his large box of spare nuts and bolts. With the repair made, we could resume motoring. We continued to drift while we had some sandwiches for lunch, and by the time we finished the wind had filled in nicely. No need to use the engine. We were able to sail all afternoon and reach Dunk Island before sunset.

Our friends on *Anwagomi* had pulled ahead and disappeared from our sight while we were incapacitated, but we had nearly caught up once we were sailing. We anchored near them in the small anchorage on the northwest corner of the island. That evening, after a good dinner on board, Dan climbed back into the aft part of the engine compartment with a flashlight and our handy-dandy magnet on a long cord to fish for the missing fourth bolt. When he brought the magnet up from the murky depths of the bilge under the engine there was the long bolt attached to it, and he replaced the short fourth bolt with this original one. We decided to wait until the next morning when we met Gordon and Miriam on shore for a hike on the island to tell them of our misadventure. They must have wondered what we had been doing all morning while they slowly sailed north.

What did we learn from this experience? Dan should have checked the coupling periodically for any loose bolts and tightened them or replaced any missing bolts. On the other hand, since the aft part of the engine is not easily accessible, and this is the first time we had a problem in over 18 years of using the engine, it didn't have a high priority. Although the chewed up plywood was unsightly, we decided it wasn't necessary to replace it, as the muffler was still well supported on the plywood shelf that was adjacent to the gouged out area. However, when we eventually sell the boat we will have some complex explaining to do for the new owners. We were fortunate that this incident happened when it did, in flat calm seas, with plenty of space to sail or drift, and no ships in the vicinity. As we often say the three components of successful cruising are the crew's skill, a good boat, and a little luck.

## Grounding in the River

Every other year since entering Australia we had sailed down the east coast of New South Wales to Sydney to escape the hot summer weather in Queensland. In the summer of 2006-2007 we wanted to explore further south to Jervis Bay. This was a large, protected bay we had visited in our small 23-foot sailboat when we lived in Sydney about 40 years before. Jervis Bay, 100 miles south of Sydney, is home to the Royal Navy College, and a small National park on the south end. Our first stop after leaving Sydney was Jibbon Beach, a small anchorage in crystal clear water in Royal National Park. As we were relaxing from our short sail, Jim and Ann, on the sailboat *Insatiable II*, arrived and anchored near by. They had sailed down from Broken Bay, North of Sydney, and were also planning to sail to Jervis Bay.

We decided to sail together, although *Insatiable II* is a much larger and faster boat than *Shaula*. They invited us aboard for happy hour and to plan our sail south. When we listened to the 5 pm weather forecast, we learned that a southerly change was coming the next morning. We decided to leave right away while we still had northerly winds behind us. The weather was perfect, with a brisk, 15- to 20-knot northeast breeze. Sailing along the coast all night, standing watches, we were just north of the entrance to Jervis Bay when the wind died at 6:30 am, and we began motoring. At 7:30 am the wind changed to southerly as predicted, but we were able to motor against it and pick up a mooring at Hole-in-the-Wall at the south end of Jervis Bay a couple of hours later. Jim and Ann, in their faster boat, had made it in by 5:30 am, just before the wind died.

Jim and Ann came over for happy hour that afternoon, and thus began two weeks of cruising in Jervis Bay in their company. We shared hiking in the nearby national park, and happy hours and dinners on each other's boats. We found a large campground ashore with wonderful hot showers which we took advantage of after our walks. Some days it poured rain, so we were confined to our boats. Jim and Ann are very gregarious, and have made many Australian friends up and down the coast. When other boats came in to the anchorage, Jim and Ann invited everyone to come over to *Insatiable II* for happy hour. For a few days Rick and Leslie on *Boomerang* joined us for hikes and happy hours. When we ran out of fresh food all three boats pulled up anchor and sailed to the west side of the bay and anchored in a small bay next to the small town of Huskisson, where we could replenish our supplies at a supermarket. While there Jim and Ann got together with an Australian couple they had met, Allen and Cindy, who were managing a hotel. They invited all of us for a barbeque at their wonderful home. The next morning we rowed over to the dock and visited the sailing school. We met Keith, the owner. We were very impressed with the magnificent clubhouse he had built with donated wood and labor. Keith teaches sailing to disabled youth in small sailboats modified for ease in handling. Jim and Ann decided to donate their old inflatable dinghy, which they had recently replaced, to the sailing school. Keith was very grateful for the gift.

From there we had a pleasant sail to the northeast side of the bay, where we planned to anchor and explore ashore, but as we approached we saw red flags up and down the beach, warning us not to anchor, as the Navy was carrying out bombing exercises. As we headed back the north winds built up, giving us a wild sail to the north end of the bay, where we re-anchored. When the wind switched to southerly, we returned to Hole-in-the-Wall on the south side of the bay.

After we left Jervis Bay we headed to the Crookhaven River, about 25 miles north. The day was beautiful and sunny, with light northeast winds. Leaving at mid-morning, we motored in flat seas and passed through the heads of the Crookhaven River at about 1 pm. To stay in the middle of the channel we were keeping a careful watch for buoys, red on the left, green on the right, as we motored up the winding river. Suddenly we were distracted by a small ultra-light plane flying overhead, the pilot sitting out in front in the open with the propeller behind him. It looked like a giant dragonfly. He must have felt free as a bird as he drifted along the river observing the beautiful surroundings. Dan and I were following his progress with binoculars, forgetting our job of staying in the center of the channel. Suddenly the boat came to an abrupt halt—we had hit a sandbank on the side of the river between two red buoys. We were hard aground. I checked the depth on our port side; it was 3.3 feet (our boat draws 5.3 feet). From past experience with going aground we knew what to do. We launched the dinghy, and Dan took our kedge anchor out in deep water about 50 feet ahead of our boat, and dropped it. When he came back aboard he cranked in the line to make it taut to hold the boat in place. A few minutes later we heard *Insatiable II* calling us on the VHF radio. They had left Jervis Bay, a couple of hours after we left, after going ashore for groceries. They asked if we were anchored, and we had to admit that we had run aground in the river and wouldn't be able to kedge off until the tide went down and rose again, probably four more hours. "Don't worry," Jim said. "We'll be able to get you off before that by pulling your boat over with our dinghy using a halyard at the top of the mast." They had helped other friends this way. We were certainly glad to have some assistance, as we were anxious to get off the sandbank and into the anchorage before sunset.

Jim and Ann are competent sailors, with years of experience cruising in Australia. They had helped us with a sail repair when the stitching disintegrated on our passage to Vanuatu. On the shore at Port Vila the four of us had spent several hours re-stitching the sail using their hand-crank sewing machine, making the sail usable for our return to Australia. We remembered that it was Jim who diagnosed our problem when our engine overheated going over the Wide Bay Bar in Australia. They have never hesitated to help us out when we were in trouble.

After our boat had sat on the sandbank for a couple of hours we were greatly relieved to see *Insatiable II* coming up the river. They motored past us to the anchorage, where they left *Insatiable*. They came back shortly in their dinghy. Dan attached a long line to the spinnaker halyard, and gave it to Jim. Jim and Ann motored out in their large dinghy with their 25-horsepower motor, pulling *Shaula* over at a 45-degree angle. We heeled alarmingly, putting the side deck underwater with seawater spilling into the cockpit. Dan was on the foredeck cranking on the anchor line. Our boat slowly inched forward on the side of the keel. After 20 minutes we could feel it lifting off the sand. I began motoring forward towards deep water. Jim released the halyard, and we were upright again. What a relief! We continued to motor up the river in the middle of the channel to the anchorage. Jim and Ann invited us over for happy hour. We thanked them over and over for their assistance. Not only do Jim and Ann provide a really sociable time, they know how to fix anything. They made our cruise to Jervis Bay and the Crookhaven River enjoyable and unforgettable.

# 12 RETURNING HOME

## Leaving Australia

After our rough passage to New Caledonia and Vanuatu in 2005, we decided we were through making offshore passages. We had to admit that as we reached our 70s, being thrown around by big seas, getting little sleep, and eating cold food when it was too rough to cook was hard on our bodies. Passage making was no longer enjoyable. Having obtained a four-year retirement visa in 2005 we had no problem with Australian Immigration. However, Australian Customs was a problem. Foreign boats are only allowed to stay in the country for two years. This time could be stretched for cruisers holding retirement visas, but not beyond three years. The only way to continue cruising in Australia was to import our boat. We didn't want to pay the high import tax because we didn't want to stay in Australia permanently. Our only options were to sail the boat home against the trade winds, or ship it home. Neither of us had any desire to sail the boat home. Having made several passages against the trades (twice from Hawaii to Seattle and once from Bora Bora to Hawaii), we knew what it is like to bash into the winds and seas day after day. We decided to have the boat shipped home.

For 14 years we had enjoyed trade-wind sailing in the warm seas of the South Pacific, where we could sail year round without suffering from cold, rain, snow, or ice. We had made a wonderful trade-wind passage from Mexico to French Polynesia. Our boat had carried us across thousands of miles of ocean, to many of the South Pacific islands, plus New Zealand and Australia. We had traveled by land on both the North and the South Islands of New Zealand, and had visited most of the major cities in Australia. Some cruisers who have sailed around the world say that the best cruising they experienced was in the South Pacific. We had enjoyed learning about other cultures, and made many cruising friends who shared our way of life and love of sailing. Altogether it had been a fantastic experience.

Leaving Australia did not mean the end of cruising for us, just a change in location. We had spent a dozen summers cruising from Seattle to Desolation Sound and beyond before we sailed to the South Pacific. We think Desolation Sound is one of the most beautiful cruising grounds in the world. It is easy sailing in mostly light winds and flat seas. The main difference is that our sailing would be limited to the summer months instead of sailing year round. After 15 years in the tropics we were spoiled; we didn't want to sail in the cold and rainy winter weather of the Pacific Northwest.

The decision of how to ship *Shaula* home was not difficult. A new company, Dockwise, had float-on/float-off service transporting yachts across oceans to desirable cruising locations. They had built several large semi-submersible ships for this purpose. This method of yacht transport didn't exist before we left for our cruise in 1993. It was a godsend for us.

## The Captain and the Admiral

Whenever we checked into a country and filled out the necessary forms Dan listed himself as the Captain of our boat, and me as crew. This seemed unfair, as we were co-owners of the boat, and had equal responsibility in its running. To our friends, Dan designated me as the Admiral, which was certainly an elevation over being just a crew member.

Although Dan had never sailed before I met him in 1965, he took to sailing readily and with enthusiasm. He turned out to be a natural-born sailor. He has always been an outdoors person. Hiking and birdwatching were important activities when he was growing up in Southern California. As a teenager he hiked extensively in the Verdugo Hills and in the San Gabriel Mountains near his home in Glendale. When he took shop in high school he worked only with hand tools, which is ideal preparation for working on a boat. After college he was drafted and served for two years in the army. He survived a typhoon on the troop ship on the way to Korea. As a forester he worked for several years as a timber cruiser for the state of Oregon, hiking in the rugged mountains. By the time he returned to school at the University of Oregon to study to become a teacher he was ready to face new challenges.

When we bought our cruising boat in 1981 Dan worked on it in the winter and spring, installing basic equipment for sailing. When we decided to sail off shore, he installed a self-steering wind vane and an autopilot to steer the boat. When our Volvo engine failed in 1987 he spent the winter installing our Yanmar with the help of a friend at Shilshole marina who is an expert. He has always worked on the engine himself, doing all the routine maintenance. During our years of cruising he was usually able to fix any equipment that broke on the boat.

When we went cruising it became evident that Dan was well suited for sailing across the ocean. He doesn't get seasick, and can fall asleep instantly in appalling conditions. His ability to tie knots and remember which knot is appropriate for different jobs is amazing to me. His woodworking skills were invaluable for making improvements on the boat. His ability to catch, haul in, and fillet all types of fish was a real asset.

My love of sailing didn't extend to cruising across oceans. First I had to overcome seasickness, which had always affected me when we encountered ocean swell crossing the Strait of Juan de Fuca.

Fortunately I learned about meclizine the year before we left on our first cruise, and found that a pill taken a half hour before departure prevented seasickness for 24 hours and had minimal side effects. I soon learned that one pill taken on the first day of a passage was all I needed. After 24 hours at sea I no longer suffered from queasiness.

After solving my seasickness problem I had to overcome my fear of ocean sailing especially at night. After crossing oceans many times without colliding with anything I realized that a collision was extremely improbable, and I stopped worrying about it. Another fear was of storms. Our boat could handle strong winds as long as we were reefed, but seas whipped up by winds were frightening. After our knockdown by a rogue wave on the way to New Zealand I became more fearful of large seas.

My other fear was of running aground, especially in the shallow waters around the coast of Australia. The thought of being stranded for hours while the tide recedes sent me into a panic. Fortunately our boat was never damaged by the sand banks on which we grounded. We were usually able to get off within a few minutes as the tide rose. Dan was never worried about going aground, but I worried whenever we were in shallow water. It seemed to us that every boat needed a worrier, and on our boat I was the designated worrier.

We both stood three-hour watches throughout our passages. In nice weather and smooth seas the watches were enjoyable, and the time passed quickly. In bad weather our watches were miserable, and seemed interminable. Sometimes we shortened the watches to one or two hours to make it more bearable. During passages Dan did all of the raising of sails and reefing and unfurling. I helped out with jobs that didn't require me to leave the cockpit, such as jibing and tacking, and pulling out or rolling in the jib. Navigation was a job for both of us. When we used a sextant on the early days of our cruising Dan took most of the sights, and we both worked out our position. When we upgraded to a GPS I put in the waypoints for our passages. It was a job I enjoyed.

My ability to tie knots was almost non-existent. Dan tried many times to teach me to tie a bowline, the basic knots for sailors, but I never mastered it. Fortunately Dan knows how to tie knots for all situations, and can learn new ones easily from a book. I did know how to sew, and had an old but sturdy sewing machine that could sew through many layers of Sunbrella. I sewed awnings and covers for everything that needed protection from the sun, and mosquito screens for all the hatches. It was my job to do the cooking, but Dan always did the dishes, which I greatly appreciated. I was responsible for provisioning the boat and kept track of the food we used so we could resupply. Dan often went with me when I shopped, and always when we did the final provisioning for a passage. When we hauled the boat out we both worked on sanding and antifouling the hull, and Dan changed the zincs and greased the seacocks. Our division of labor worked well for us. Neither of us complained about our jobs on the boat. With all the work that cruising entails, and the necessity for being alert for ships and changes of weather while underway, we couldn't imagine how single-handed sailors could manage long passages.

For me the passage making was merely the means to an end. What I enjoyed was traveling to new places and learning about how people live in cultures much different from ours. Socializing with other cruisers from all over the world was another big attraction. I was willing to put up with the discomfort and sleeplessness during a passage in order to reach the wonderful destinations.

It is a wonder that Dan and I ended up as long-distance cruisers. I grew up in Knoxville, Tennessee, far from the ocean. Dan grew up in a suburb of Los Angeles, also distant from the ocean. Neither of us had relatives or friends who had ever sailed across the ocean. I loved sailing from the beginning, and after I taught Dan to sail he shared my love of sailing. It was Dan's avid reading of sailing adventure stories in books and magazines that eventually led to our cruising across the Pacific Ocean. It turned out to be an adventure of a lifetime.

## Our Ideal Cruising Boat

Our boat, a 28-foot Bristol Channel Cutter, had a major role in making our cruising successful. We both loved Bristol Channel Cutters from the first time we sailed one on Puget Sound. It is a small fiberglass version of classic wooden boats that served as pilot boats off the coast of England. These boats were built to withstand all kinds of weather, and have proved to be both seaworthy and sea kindly. As a cutter it has a flexible sail plan, making it possible to sail in light winds using all three sails or in strong winds to sail with just a reefed mainsail. Although its length on deck is only 28 feet, because of its seven-foot bowsprit it carries the sail area of a 35-foot boat, and as a result it sails well in light winds. We can often pass sailboats that are much longer. We were impressed with its strength when we were hit by a rogue wave in the Tasman Sea on the way to New Zealand. The force of tons of water breaking over our boat broke our boom in two, but the boat itself remained intact. There were no cracks in the hull, no broken ports, and no hatches torn off. Our little boat proved it could stand up to heavy weather.

The boat is designed for two people to cruise, but because of its small size one person can easily sail it. Its small cockpit is an advantage in heavy seas. When waves break into the cockpit they end up in the small cockpit well, where the water is quickly drained back into the ocean through two hoses in the bottom of the well. It is safer down below in our boat than in a larger boat because we can hold onto railings on both sides of the boat at the same time, so there is less chance of being thrown across the cabin. For its size it has a huge amount of storage space. We can easily provision with six months of staples, and we have room for hundreds of books.

Because of its classic design, with its pleasing amount of shear and its sturdy rig, we get many compliments on our boat. It has a well-deserved reputation for being an excellent cruising sailboat. Many people have told us that a Bristol Channel Cutter was their dreamboat. After our cruises across the Pacific, learning how well it sails in a variety of conditions, we have great confidence in our boat and appreciate its qualities even more.

During our years at sea there have been many improvements in cruising equipment, making long-distance sailing easier. We kept up with most of the changes, installing new equipment and upgrading as we went along. On our first ocean passage to Hawaii in 1985 we used a sextant for navigation, but three years later, when we sailed to French Polynesia, Hawaii, and back, we used a Sat Nav system (a precursor to GPS). By 1993, when we began our 15-year cruise, we were able to buy a GPS at a reasonable price. Now nearly all sailors navigate using GPS, and it has made crossing oceans and cruising in general much easier and safer.

For communication we started out in 1985 with a ham radio which could also be used as a marine single side band radio. Dan used Morse code to communicate with his father, although he used voice to talk on the ham net. In later years we used both single side band and ham radio to talk to other cruisers making passages. During our cruise from 1988 to 1990 we wrote our newsletters by hand, and then mailed them to my sister in Alabama, who copied and mailed them to our friends and relatives. When we left in 1993 we had a laptop computer on board, and used that for writing newsletters, which was much easier than writing them out longhand. In the late 1990s, it became possible to use the ham radio hooked up to a computer via a modem to send and receive email on the boat. Now we could email our newsletters to my sister as an attachment. It was also possible to send and receive email from friends and relatives on land when we were in the middle of the ocean. A bonus was that we could receive detailed up-to-date weather maps on the computer while underway. Having email on the boat was a tremendous improvement in communications.

When we reached New Zealand we purchased and installed a radar unit on our mast. This makes it safer to sail in conditions of poor visibility and helps in determining whether a ship is on a collision course with our boat. As solar panels became more efficient and cheaper we upgraded ours, starting with two small solar panels in 1988, then adding two more later. Finally we bought two large, more efficient solar panels and installed them on both sides of the boat on brackets that allow us to orient them in all directions to face the sun. The other big improvement to our boat was installing a wind generator. In the moderate to strong winds off the coast of Australia we were able to get lots of power from the wind, and combined with our solar power we had excess power for keeping our batteries charged up. Because of this excess power we were able to install a small portable electric refrigerator. This was a big improvement over keeping food cold in a small ice chest which required buying a block of ice every two or three days when in port. We seldom had to run the engine to charge the batteries. We felt good about having a very small carbon footprint during our 15 years of cruising.

During our cruising years we replaced or upgraded much of the equipment with which we started our voyage. We replaced our old analog depth sounder with a modern digital fish finder, and our small manual anchor windlass with a faster, larger two-speed manual windlass. We eventually replaced our small handheld GPS with a larger unit with numbers large enough to read from several feet away. We've replaced some of our incandescent reading lamps with LED lights, which use much less power.

Our dodger eventually wore out, so we had a new one made in Brisbane out of Stamoid, a more durable material than Sunbrella. Because of the intense sun in Australia we had a bimini made to shade the cockpit. We can leave it up while sailing and still have good visibility over the top of the dodger. It is also useful for keeping us dry when we sail in the rain. I made shade awnings for the foredeck and cabin, and replaced the fore hatch cover several times after it degraded in the sun. We gradually painted over most of the varnished wood on deck, making maintenance much easier. Varnish lasts only a few months in the tropics, while paint lasts for three or four years.

Dan has improved the interior of our boat in numerous ways. He installed fold-down shelves on the bulkheads on both sides of the boat, added compartments for the pots and pans locker, put hinges on the ice chest lid for easier access, and added a removable shelf in the galley. In 2003 he moved the house batteries from the starboard side of the boat to a battery box in the center under our ladder to get rid of our starboard list, and upgraded our electrical system.

All of these improvements have made life on our boat better. Our boat shows little wear and tear after 15 years of cruising and living aboard, and we enjoy it even more for sailing in local waters and to Desolation Sound. We hope to sail in local waters for many more years. With our self-steering wind vane, solar panels, and wind-generator our boat looks like a world cruiser. We still get many compliments on our boat, and sailors often ask us where we have cruised. Many people we meet have ambitions to cruise in the future or have returned from cruising long distances. We love sharing the stories of our cruising with other sailors, and learning about other people's adventures.

## Shipping *Shaula* Home

Dockwise service didn't exist when we started our cruise in 1993. For us it seemed the ideal way to transport our boat from Brisbane back to Seattle. Our boat would be securely propped up on the deck of the large Dockwise ship, along with dozens of boats of all sizes, and transported out of the water over 7,000 miles from Brisbane to Ensenada, Mexico. It would save wear and tear on our boat and on us. The estimated cost of $25,000 was a shock. The cost is based on the amount of space your boat will take up on the deck of the ship. By removing our bowsprit and self-steering gear we could reduce the 38 foot length to 31 feet, and save thousands of dollars. Dockwise gives a 20 per cent discount for customers who pay five months in advance. This would reduce our cost to about $19,000. The next departure was in January 2008, so we reserved a space and paid for it in August 2007.

Once we had made our reservation we did not want to stray too far from Brisbane, so we decided to visit Great Keppel Island, a two-day sail north of Brisbane. Great Keppel Island is an ideal cruising area. There are protected anchorages on the north, west, and south sides of the island. For six weeks we enjoyed hiking the many trails crossing the island, and swimming in the warm swimming pool at the resort, and in the saltwater at the anchorages. In a few hours we could sail from the island to Keppel Bay Marina on the mainland to resupply with fresh food at the nearby town of Yeppoon. Great Keppel Island is a favorite with Australian cruising sailors. We enjoyed socializing with other cruisers and relaxing. On our last day there we were anchored at Long Beach, at the south end of the island, and shared a happy hour ashore with the crews of 25 other boats at anchor. It was a memorable occasion.

In November we sailed back south to Bundaberg, where we enjoyed visiting old friends at the marina. In December we motored down to

# RETURNING HOME

Mooloolaba, just north of Brisbane. Friends Betty and Steve on *Jams* were staying at the marina, and we enjoyed visiting them and Jim and Ann on *Insatiable* while we were there.

A few weeks before Dockwise was to depart we sailed down to Brisbane in order to get our boat ready for shipping. At the end of January 2008 the large Dockwise ship, *Super Servant 4*, came into harbor partway up the Brisbane River. We motored up the Brisbane River and stayed in the marina adjacent to the loading dock. It was a day of frantic activity, as we prepared the boat for the passage. We removed the sails and everything that was tied down on deck and put it below, covered the dodger with a tarp to protect it from the ship's exhaust, and packed our clothes for the plane trip to Los Angeles. The next morning the crew of *Super Servant 4* flooded the ship's ballast tanks with millions of gallons of water to partially submerge it. They announced the order in which we would board. Boat owners would then motor their boat onto the stern of the Dockwise ship and tie up at the designated spot on the deck, We milled around the bay along with all of the other boats listening on the VHF radio as each boat name was called. When our boat name was called we motored onto the ship to the designated spot and tied up to the boat next to us that was tied to the rail. We locked our boat, and then walked the length of the deck on a narrow catwalk to the bow of the ship, signed the necessary papers, and departed.

Our friends, Bob and Judy from *Siddiqi*, had offered to let us stay with them at Judy's parent's house just north of Brisbane after our boat was loaded onto the ship. Our plane reservations were for two days later. It was great to have a place to stay and a last visit with Bob and Judy and Judy's parents before we departed. The boats would all be transported to their destinations, first to New Zealand, then on to Ensenada. It would take about 30 days to reach Mexico. All we had to do was wait for an email to tell us when Dockwise would arrive in

Ensenada, and be there on the day of arrival. The Dockwise ship would submerge, and we would motor our boat off and into the nearby marina. It couldn't be easier, and saved us at least six months of arduous sailing across the Pacific.

The 30 days our boat was being transported across the Pacific went quickly. We had a nice visit with Dan's mother in her retirement home in Glendale. After two weeks we rented a car and drove to Palm Springs to meet our friends Gordon and Miriam. They had invited us to visit them at their home in Joshua Tree. While there we were able to take hikes on the many trails in nearby Joshua Tree National Park. When we were notified of the date of Dockwise ship's arrival Gordon and Miriam drove us down to Ensenada, Mexico. They had shipped their boat by Dockwise the year before, and were working on it at the marina to prepare it for cruising in Mexico.

Our boat arrived in Ensenada in good shape, although we noticed salt crystals 35 or 40 feet up towards the top of the mast. The ship must have gone through some rough weather to have spray reach that high in our rigging. We were glad we had not crossed the Pacific in our small boat. After a couple of weeks in the marina in Ensenada we sailed on an overnight trip to San Diego, where we stayed several weeks while waiting to have the boat trucked to Olympia. Having a boat trucked is much harder than having it shipped by Dockwise. It involves removing the mast, wrapping it in plastic sheeting to protect it, stowing everything else inside the boat, and finally having the boat hauled out of the water and onto the truck. While the truck transported our boat up I-5 to Olympia we took Amtrak from San Diego. We stayed with our friends Jim and Lynn in Olympia while we worked on *Shaula* in the boatyard. For several weeks we cleaned the salt crystals from the rigging, cleaned the topsides from all the road grime, anti-fouled the bottom, and had the mast put back on the boat. Then we had a short sail from Olympia to Seattle.

# RETURNING HOME

Now that we had our boat back home we had to learn to live on land again. It was like starting over, a new phase in our lives. For the first time we would be retired while living in Seattle. We were able to prolong living aboard by sailing to Desolation Sound for the summer. Finally, eight months after leaving Australia, it was time to buy a car, rent an apartment, put our boat in a slip at Shilshole Bay Marina, and buy some warm clothes so we could survive the winter weather. We took our 58 boxes of belongings out of storage, and bought some furniture from craigslist to furnish our one-bedroom apartment. The apartment in Ballard was only 640 square feet, but seemed enormous after living on a 28-foot boat for 15 years.

One thing we realized immediately was that it was much more expensive to live on land than on our boat. Seattle had changed considerably in the 15 years since we'd lived here. We were disappointed to see small homes being torn down to build six-story apartments and condos in Ballard, but we liked the new Benaroya Hall, the striking new downtown and Ballard branch libraries, and the renovated marina at Shilshole. It is wonderful having the boat so close that we can maintain it in the winter. In the summer we enjoy sailing in *Shaula* to Desolation Sound. We still get together with cruising friends who have also returned from offshore cruising to reminisce about the good old days cruising in the South Pacific.

# ACKNOWLEDGEMENTS

I would like to thank the many friends who encouraged me to write the story or our sailing adventures.

I especially want to thank Claire Anderson and the members of the Ballard NW Senior Center memoir class. They listened to my stories and provided many helpful suggestions for improvements. Claire indulged my not writing on her suggested topic each week, but allowed me to write whatever I wished. I appreciate her advice in improving my writing, and her encouragement and guidance in the publishing of this book.

I am grateful to my copy editor, Janet Southcott, for her expertise and care in polishing my manuscript.

My heartfelt thanks go to my husband Dan Dews, who shared this adventure. His enthusiasm for ocean sailing and his ability to fix anything that broke on the boat made the voyage possible and enjoyable. I thank him for editing my stories before I read them in the memoir class and for his patience and help during the preparation of this book for publication.

ABOUT THE AUTHOR

Alice Wiersema was born in Knoxville, Tennessee. She obtained her B.S. and M.S. in Chemistry at the University of Illinois and the University of California, Berkeley. Living in Berkeley gave her the opportunity to learn to sail, and she soon became an avid sailor. After working as a chemist in Chicago and California she enrolled in graduate school at the University of Oregon to study biology. She married Dan Dews in Eugene, Oregon and they settled in Seattle, where Alice worked as a Research Technician and Dan taught high school science. While working they spent their summers sailing locally in small boats. After lifetimes in science and education, Alice and Dan took early retirement and spent the next fifteen years cruising across the Pacific in their twenty-eight foot sailboat, *Shaula*. Since returning to Seattle Alice has resumed her musical interests in violin and singing. The couple continues to sail each summer in *Shaula* in the protected waters of Puget Sound and British Columbia.